£15

CW01678122

RAF OPERATIONS 1918-38

By the same author
Calshot, 1913-61
The Flying Elephants – 27 Sqn RFC/RAF
Mosquito at War
Hurricane at War
Airmen of World War One
Sunderland at War
Hampden Special
Beaufighter at War
Path Finders at War
Albert Ball, VC
For Valour – The Air VCs
Sopwith Camel – King of Combat
History of the RAF
Guns in the Sky – The AGs
Coastal Command at War
Spitfire
Fighter Command, 1936-68
Age of the Biplane
Air War over Europe, 1939-45
Desert Air Force at War (co-author)
Encyclopaedia of British Military Aircraft
Bomber Group at War
Wellington at War
Images of Air war, 1939-45
Bomber Barons
Eugene Esmonde, VC, DSO
Royal Air Force Handbook, 1939-45
Bristol Blenheim
Fighter Pilots of the RAF, 1939-45
Mosquito Squadrons of the RAF
Men of the Desert Air Force, 1940-43
Gloster Meteor
Men of Coastal Command, 1939-45
Bristol Fighter – King of Two-seaters
Tales from the Bombers
The Wellington Bomber
Beaufighter

Edited:
Bomber Pilot, 1916-18
Fighter Pilot on the Western Front
Wings over the Somme
Fall of an Eagle – Ernst Udet

RAF Operations
1918-38

CHAZ BOWYER

WILLIAM KIMBER · LONDON

First published in 1988 by
WILLIAM KIMBER & CO. LIMITED
100 Jermyn Street, London SW1Y 6EE

© Chaz Bowyer, 1988

ISBN 0-7183-0671-6

This book is copyright. No part of it may be reproduced in any form without permission in writing from the publishers except by a reviewer who wishes to quote brief passages in a review written for inclusion in a newspaper, magazine, radio or television broadcast.

Photoset in North Wales by
Derek Doyle & Associates Mold, Clwyd
and printed in Great Britain by
Redwood Burn Limited
Trowbridge, Wiltshire

Contents

List of Illustrations

Acknowledgments

While I am indebted to many people over past years for their selfless material aid and unceasing co-operation in my researches into RAF history, particularly in the context of photographs and original documentation, I wish to record my special gratitude to the following for direct assistance in the compilation of this volume, listed in no significant order of precedence: Bruce Robertson, Squadron Leader R.C.B. 'Chris' Ashworth, RAF Retd, Flight Lieutenant 'Andy' Thomas, RAF, Ray Vann, and the late and much missed Christopher C.H. Cole, Ray C. Sturtivant, G. Stuart Leslie, Mrs D. Cowton, No 31 Squadron Association, 'Reg' Mack of the Department of Aviation Records, RAF Museum, Hendon, Wing Commander C. G. Jefford, MBE,RAF and never least Dave Gray and Jane Richardson of Walkers Studios, Scarborough.

Introduction

On 1 April 1988 the Royal Air Force celebrated the 70th anniversary of its original formation. Conceived and born in the throes of a bloody European war, the RAF is probably best remembered in the lay mind today for its vital contributions during the global war of 1939-45 and various post-1945 activities. Yet it is rarely realised that during its first twenty 'adolescent' years, 1918-38, the RAF had also been committed to full warlike operations on virtually *every* day of those so-termed 'peace' years *somewhere* around the globe. Such operations varied from local policing skirmishes to near full-scale wars involving many thousands of British servicemen. And lest the reader looks askance at my deliberate use of the description 'war', I would point out that between May 1920 and December 1938 no less than five Victoria Crosses – three of these posthumous – were awarded to Army personnel alone during hostilities in Mesopotamia and India, apart from a host of lesser gallantry awards; operations in which the 'local' RAF was unfailingly involved.

During the years 1918-38 the infant RAF, fiercely defended and championed by Hugh Trenchard, John Salmond and other far-sighted visionaries, slowly matured despite outdated thinking at brasshat levels in other Services, successive misguided and unsympathetic political masters, parsimonious financing, governmental 'conveniences', and, not least, actual conditions of service within the RAF which might well have daunted lesser men. Moreover, it should be well marked that every man and boy in RAF uniform during those years was an unpressed volunteer for his chosen career. In constant jeopardy of dissolution from the machinations of army, naval and political enemies in high offices during its early years, the

RAF gradually established itself alongside its elders, the Army and Royal Navy, as an equal partner – albeit still very much a 'Cinderella squashed between two great Ugly Sisters', as one contemporary wit dubbed the relationship. Those two decades saw the 'junior Service' (*sic*) learn fundamental lessons of sharp end operational employment of air power in the toughest school of all – sheer experience – thereby extending and expanding its existing limited knowledge gained bloodily during the latter stages of the 1914-18 war. That ingrained experience, allied to a maturing realisation of the huge potential of air power, was acquired by many relatively junior officers who later, as senior commanders, were to lead and direct RAF operations and policies during 1939-45 and the years beyond.

The prime purpose of this book is simply to *begin* to fill the gaps in published records of the RAF's *operational* activities during the years 1918-38, culminating with the notorious Munich Crisis of September 1938. Accordingly, I have omitted references to such well-publicised peaceful activities as the annual Hendon Air Displays, sham war exercises, Schneider Trophy races, and similar UK-based facets of RAF life then. One other major, but deliberate, omission concerns the unpublicised yet valuable contributions of the Fleet Air Arm. Although legally under Air Ministry aegis for most of those years, and indeed generally regarded by many then as merely one more facet of RAF organisation, the Fleet Air Arm, in my view, was always in essence a separate formation with esoteric duties, methods, and generic character, demanding a volume to itself if it is to receive worthy recognition.

Patently it is not possible to detail *every* single operational undertaking by the RAF during those years within the covers of one modest volume, hence I have simply attempted here to paint a broad canvas of the more significant events upon which further details may readily be added by those readers seeking deeper knowledge and understanding. Nevertheless, I have intentionally included what might be termed background events and circumstances where the RAF was eventually called upon to take direct actions in order to provide the contemporary atmosphere in each situation.

Above all, it should be remembered that all the events described in the following pages took place in an era when Britain still possessed a vast global empire, with additional legal responsibilities for the control and/or governing of other foreign mandated territories. That era of imperial glory and responsibility is now long past and too often misunderstood or even forgotten by today's generations. Perhaps such generations should recall the words of Cervantes:

'History is a depository of great actions, the witness of what is past, the example and instructor of the present, and the monitor of the future ...'

Chaz Bowyer
Norwich, 1988

CHAPTER ONE

Background

When the Royal Air Force came into being on 1 April 1918, resulting from an amalgamation of the former Royal Flying Corps and Royal Naval Air Service, it was the world's first independent (of Army and Navy control) air service. Seven and a half months later, on 11 November 1918, when an armistice agreement brought a halt to a semi-global war of a scale never before witnessed in human history, the RAF was also the largest air force in the world. On that date RAF personnel strength comprised more than 27,000 officers and nearly 300,000 non-commissioned airmen and airwomen.* These men and women manned totals of 188 firstline squadrons, 187 training squadrons, 15 separate Flights, and numerous non-flying units, while overall the RAF could count a stock of 3,300 aircraft and 103 airships. In addition several thousands more aircraft intended for RAF use were crowding the assembly lines and factory floors of a host of private or government-subsidised aircraft manufacturers throughout Britain.

The cessation of hostilities in November 1918 brought an immediate, albeit natural, reaction to the years of slaughter as the British nation turned its thoughts to the immediate future of peace. Equally naturally the vast majority of men in the armed Services, committed in the main to 'hostilities only' engagements, opted to return to civilian life in order to resume or commence future careers. Demobilisation of all three armed Services was rapid, even hasty, and it must be said badly organised, but the results literally decimated the RAF in particular. By 3 January 1920 totals of 26,087 officers, 21,259

* On 1 April 1918 the Women's Royal Air Force (WRAF) was also officially created, but was disbanded on 1 April 1920, by which date approximately 32,000 women had served in its ranks.

cadets, and 227,229 other non-commissioned ranks had been discharged from RAF service. Of the 99 firstline squadrons based along the Western Front in Europe in November 1918, only five still existed by October 1919 and even these had dwindled to a single squadron, No 12, by the end of that month.

On 11 January 1919*, Hugh Trenchard was re-appointed Chief of the Air Staff (CAS), and in the same month Winston Churchill was appointed by Lloyd George as the dual Secretary of State for War *and* Air. The latter's anomalous appointment was accompanied by Lloyd George's stated intention to abolish the Air Ministry as a separate department, although by December 1919 the Lloyd George Government came to have second thoughts on the matter. Churchill, undecided as yet what to actually do about the Air Department, consulted with his predecessor at the Air Ministry, Lord Weir, who took the opportunity to advocate the re-appointment of Trenchard as CAS, insisting, 'Trenchard's your best man. He can make do with little and won't have to be carried.' Accordingly, Churchill invited Trenchard to meet him at the War Office in early February, offered him the post of CAS, then asked Trenchard to outline his personal ideas on paper for reorganising the Air Ministry and, more crucially, the basis for any future RAF.

Trenchard's task, to rebuild an air force from the ashes of its former giant strength, was daunting. Restricted severely financially by a Cabinet decision (later) to limit Air Estimates to £15 million per annum for the next five years, and in the face of violent opposition by both War Office and Admiralty hierarchies (and no few senior politicians) pressing constantly to dissolve the infant RAF and split it into purely army and navy-controlled elements, Hugh Trenchard commented: 'I was left with nothing but two heaps of rubble – one of bricks and mortar, the other of men. There were gems in each heap, but I had to pick them out blindfold.'

Further obstacles to establishing a permanent future for the RAF followed quickly in 1919, not least of these being the Ten

* A retrospective date of official appointment.

Year Rule instigated by Lloyd George and endorsed by the full Cabinet, based on the premise that Britain would not need to fight any major war for a decade; a decision which effectively tightened the Treasury's hold on all three Services' financial budgeting. With an Air Estimate already merely one-eighth of that allotted to the army, the RAF became the most vulnerable target for any future restrictions in public expenditure.

Yet one more frustration for Trenchard was the growing intransigence of the restless Churchill who began leaning towards the views of the RAF's opponents in high office. Trenchard, realising that his hopes for the RAF were rapidly reaching crisis point, characteristically tackled Churchill face to face on the subject, having a blazing row in the Minister's private office with no verbal holds barred.* When tempers had cooled and apologies had been exchanged, Churchill asked Trenchard to prepare a written statement of his ideas which Trenchard immediately drafted that same day. About 3,000 words in length, and dated 11 September 1919, Trenchard's paper succinctly summarised the necessity for the RAF to be a separate Service for (primarily) future strategic reasons, but at the same time emphasised that the proposed foundations for the RAF could be accomplished economically.

By December 1919 Trenchard's proposals had been prepared as a White Paper and – titled colloquially as 'Trenchard's Memorandum' – was introduced in Westminster by Churchill on 11 December 1919. The reaction from most politicians, Service chiefs, and Fleet Street press barons was either apathy or hostility, particularly in the matter of the RAF becoming a 'Third Service' unshackled by the Admiralty or the War Office. The principal points of the 'Memorandum' concerned literal foundations – the 'bricks and mortar' as Trenchard had called them and to which he proposed to allot thirteen of the fifteen million pounds' budget initially – and these included an RAF College for officer-cadets at Cranwell, an Air Staff College for senior officers at Andover, and a

* Prior to this meeting Trenchard had actually prepared his resignation as CAS as a 'test' of Churchill's faith in the RAF, but tore this up privately next day before going to see the Minister.

future RAF Aircraft Apprentice school of technical training at Halton.

Other proposals, previously submitted with Churchill's approval, included the introduction of the Short Service Commission scheme of officer enrolment, whereby some two-thirds of future officers would be offered engagements of three, four or five years service, with a further four years' Reserve commitment, and limited opportunities for permanent commissions; a proposal unprecedented in the context of Britain's armed Services but one which would ensure Trenchard's vision of his future RAF as a 'young man's service' with ample scope for more rapid promotion prospects for permanently commissioned officers than was the case in the tradition-bound army and navy ranks. Overall, Trenchard's prime concern was to set and maintain the highest possible standards of training and performance in all spheres of RAF life; a maxim pursued by the RAF thereafter and until the present day.

Of the available 25½ squadrons Trenchard proposed to base no less than 19 of these overseas, as follows:

India	8 squadrons and one depot
Egypt	7 squadrons and one depot
Mesopotamia	3 squadrons and one depot

plus detached Flights of seaplanes based at Malta, Alexandria, and carrier-borne in the Mediterranean. The natural headquarters for all these units in the so-termed Near and Middle East zones was thought to be best located in Egypt, in itself also a vital junction in the projected civil Imperial Air Route to India. Questioned in the Commons on 15 December 1919 about this decision to base the bulk of the RAF abroad, Churchill reminded the House that Britain still had a vast empire and other countries to administer and control, then stated, 'The first duty of the Royal Air Force is to garrison the British Empire.' Although Winston Churchill had been Trenchard's support in certain ways during the latter's struggles to preserve an intact RAF during 1919 especially, his mercurial nature gave Trenchard many anxious moments. In Trenchard's own words:

> The same Winston Churchill caused me more sleepless nights while he was in charge of the Air Ministry than any of his successors. His imagination was too strong for comfort and he tended to be too easily swayed by the last devil's advocate he happened to meet. The trouble was that the devil then had too many advocates.*

Churchill finally ceased to be Secretary of State for Air on 5 April 1921, being succeeded in that appointment by Captain Frederick Guest.

Trenchard's unceasing defence of 'his' RAF against the traditionalist, die-hard army, navy, and political would-be agents of destruction was to continue throughout his lengthy tenure as CAS, and in lesser degree for the following decade. Meantime, the overseas elements of the RAF steadily justified their existence in a progression of operational employments, as will be described in subsequent chapters. In all of these Trenchard's prime contention that a few RAF squadrons were perfectly capable of 'controlling' vast areas of desert and mountain kingdoms without the need for physical occupation of such territories by the traditional, cumbrous – and far more expensive – army garrisons began to be demonstrated with undoubted success. Due to continuing restrictive financial budgets the RAF remained small in overall strength – the year 1920's budget allowed for personnel strengths of 3,280 officers and 25,000 non-commissioned ranks, for example – but Trenchard's creed of laying firm foundations was soon to bear fruit. The RAF College at Cranwell was 'opened' on 5 February 1920, the Air Staff College at Andover on 4 April 1922, while the Aircraft Apprentice School at Halton had commenced in being in 1920, albeit lacking complete accommodation for all entrants to the scheme until 1926.

It should be added that initially Trenchard allocated the

* Quoted in *Trenchard – Man of Vision* by A. Boyle; Collins, 1962

FATHER OF THE RAF. MRAF Lord Hugh Trenchard, Chief of the Air Staff, 1918-29.

major portion of his budget buildings fund to Halton rather than Cranwell or Andover, emphasising thereby his conviction that the future RAF would rely heavily upon its engineering personnel to maintain the high standards of performance he never failed to propound. As with almost all his 'visions', 'Trenchard's Brats' – as the products of Halton and other boy training establishments came to be dubbed by direct entry airmen in the years to come – fulfilled all Trenchard's hopes for his scheme. Over the next 60 years ex-Brats provided the technical backbone to the RAF, with a host of such men rising to very senior ranks, including one Chief of Air Staff.

Trenchard's ploy of basing the bulk of the RAF's firstline units outside the United Kingdom found an early confirmation of the efficacy of air power and potential in the brief campaign against the 'Mad Mullah' in Somaliland in early 1920, and this success provided an object lesson for future employment of overseas RAF units. It also gave Churchill, recently appointed Secretary of State for the Colonial Office in March 1921, a slim basis for supporting Trenchard in his views that the RAF could become a major instrument of peaceful control in such troubled areas as Palestine and Mesopotamia – territories which were mandated responsibilities of the British government until such countries had been prepared for self-government independently. With the creation of a new Middle East department within the aegis of the Colonial Office from 1 March 1921 – thereby bringing all matters concerning Palestine, Transjordan and Mesopotamia together under Churchill's hand – Churchill called Trenchard to accompany him to a conference in Cairo, Egypt in March 1921 where, brooking no opposition to his personal intentions, Churchill imposed the principle of RAF squadrons substituting for large army contingents in the Middle East mandated territories, with particular emphasis on such 'air control' in the case of Iraq* where the principle of gradual shifting of military control from the army to the RAF was to be instigated by late 1922. This principle of air control, though grudgingly accepted by

* Mesopotamia was retitled Iraq in September 1921

Parliament in the summer of 1921, was not without its vociferous opponents and detractors in influential political and Service circles, but despite such criticism and, indeed, deliberate obstruction and lack of co-operation from the War Office in particular, Trenchard continued to plan for the future, basing his forward planning on thrift and self-sufficiency for the RAF.

Trenchard's specific proposal for air control in Iraq was for eight squadrons to be based there. Of these five – Nos 6, 8, 30, 55, 84 – were already in the area, and the remaining three – Nos 1, 45 and 70 – were transferred by early 1922. Actual control vested in an RAF officer was finally assumed in Iraq on 1 October 1922 when Air Vice-Marshal John Salmond took up his new appointment as GOC *all* forces in the theatre. By then the RAF had already firmly established its ability to quell lawlessness among the Arab tribes merely by its presence and immediate reactions to any specific uprising or unrest. This deterrent effect was accomplished often merely by demonstration flights without any necessity for using bombs or guns – the sheer psychological fear of near-instant retribution from the skies being sufficient to impress potential or actual rebels.

Trenchard's own orders were strict in such affairs; Arab rebel villages *et al* were never to be actually bombed or strafed until the inhabitants had received at least 24 hours' notice (by air-dropped leaflets) of any punitive raid.* To his Middle East RAF Commander Trenchard laid down three simple principles to be followed:

> The air force is a preventative against risings more than a means of putting them down. Concentration is the first essential. Continuous demonstration is the second essential. And when punishment is intended, the punishment must be severe, continuous, and even prolonged ... †

* Trenchard also, initially, forbade the carrying by RAF air crews of 'safety chits' promising Arabs rewards for safe conduct of the crews in the event of forced landings among tribal areas. Later such chits were to be commonly carried by all air crews.

† Quoted in *Trenchard – Man of Vision* by A. Boyle: Collins, 1962

Not that the RAF's sole preoccupation in Iraq concerned bombing; on several occasions individual RAF crews, on their own responsibility, used their aircraft to convey seriously sick Bedouin sheikhs to Service hospitals for treatments which saved the Arabs' lives – mercy missions undertaken quite spontaneously which impressed the tribes greatly.

Although the implementation of Trenchard's policy for air control in Iraq quickly proved highly successful, the opposition to the RAF as a separate Service from the army and navy hierarchy – in particular Admiral Lord Beatty – was far from diminished within the Whitehall corridors of political power. The fate of the RAF remained balanced on a knife-edge for several years as succeeding governments of differing political shades became responsible for decision-making in this context. By the mid-1920s Trenchard had become well-used to the varying methods – both open and underhand – employed by opponents of the RAF and continued skilfully to ward off or scotch each attempt to dissolve the Third Service, while still patiently building for the future in both material and policy veins.

Ever conscious of the vital need to limit his financial demands for the RAF – economic grounds being his lynchpin for retaining his air force as a distinct entity – Trenchard demonstrated yet another facet of his immense vision by proposing the creation of the Auxiliary Air Force (AAF) – a 'Weekend Air Force' – as a Territorial force; a scheme provided for in both the original Air Force Constitution Act of 1917, and in Trenchard's 1919 Memorandum, but which did not become law until 1924, with the first AAF squadron (No 602) being officially formed in 1925 – a scheme, it might be added, that Churchill dogmatically opposed for reasons known only to himself.

By 1925 Trenchard's unceasing fight to retain the RAF as a service in its own right was largely won – though by no means entirely over – and he could expend more time on his many patient proposals for a steady expansion of RAF firstline squadrons and their necessary ancillary equipment in men and materials. In that same year Winston Churchill, after changing his political allegiance, and recently appointed as Chancellor of the Exchequer in Baldwin's government, launched his first Budget on 28 April with a prime theme of a return to the gold standard which had been suspended since the 1914-18 war.

This 'dear money' policy was to have grave results overall, and in the context of the armed Services soon brought a reduction in allotted finance by some three per cent initially; a decision which immediately put a brake on RAF expansion and revived the inter-Service rivalry as each sought a maximum share of any available budget. Even the old arguments – instigated primarily by Beatty at the Admiralty – to hive off part, if not all the RAF to army or navy control rose to the surface yet again, part-proposed by Churchill '... in the interests of economy' (*sic*). Against a backcloth of burgeoning civil unrest on the domestic front and an increasing near-hysterical international movement towards total disarmament among many European nations, RAF expansion during 1926-28 virtually ground to a halt.

Then, in 1928, Churchill initiated an unending extension to the Ten Year Rule originated in 1919 by Lloyd George; a patent excuse for restricting finance for the armed Services which was not repealed until 1932 and even then did not result in any increased budgets for the Services until 1934. As to the existing budgeting for the RAF, J.L. Garvin, editor of *The Observer*, in March 1929 stated that of the near-£600-million spent on all three armed Services during the previous five years, on average a mere six per cent had been allotted to the RAF.

On 1 January 1930 Hugh Trenchard finally relinquished his task as Chief of the Air Staff, being succeeded in his post by Air Chief Marshal Sir John Salmond on that same date. It had always been his intention to 'make way for a younger man' once he felt that the RAF had secured a permanent future as its own master – indeed, his formal resignation as CAS had first been submitted twelve months before it was actually accepted and formalised – although in accordance with the contemporary King's Rules and Regulations (KRR) as a Marshal of the RAF he remained on the Active (half-pay) List of the RAF.

Despite the continuing miserly financing for the RAF at the moment of his retirement, Trenchard had at least the private satisfaction of having watched the blossoming of the many basic 'foundation stones' he had 'laid' in 1919. The Cadet College at Cranwell and Aircraft Apprentice School at Halton had already begun to pass graduates into the ranks of the RAF; the Auxiliary Air Force had quickly proved entirely successful; the Short Service Commission Scheme had proved its worth despite its many detractors. The actual squadron strength of the RAF had slowly grown numerically, though was still very far short of Trenchard's many proposals and succeeding governmental promises. The actual equipment of those squadrons throughout Trenchard's tenure as CAS was in the main vintage 1914-18 aircraft, with similarly aged weaponry. By 1930 the stalwart De Havilland 9A had only begun to be retired from operational use, but its doughty companion, the Bristol F2B, was still in firstline squadron use overseas.

Such a state of affairs was hardly Trenchard's fault; his niggardly allotted annual budgets for the RAF had left him nothing to spare for proper modernisation of his Service ironmongery. Dubbed 'Father of the RAF' – a title he detested personally – Hugh Trenchard's vision and moral courage had created, nurtured, championed, and jealously shielded 'his' infant RAF until it was capable of standing on its own feet as the nation's prime future protector in peace and war. Without his stubborn conviction, immense faith, and unceasing labours, the Royal Air Force would certainly have been dismembered in infancy. And had his opponents in the 1920s achieved that goal, the course of the war of 1939-45 would equally certainly have taken a far different direction. Above all, perhaps, Trenchard, in part due to the limitations of his budgets but primarily to his deliberate intention, had established an over-riding principle for the RAF of 'quality before quantity' in every facet of the adolescent Service. His emphasis on highest possible standards in all aspects of training and later application gave the RAF a rock foundation upon which to eventually expand; a principle which the RAF has adhered to ever since.

His successor as CAS, John Salmond, though of an entirely different character, was an equally fierce champion of air power and the necessity for keeping the RAF as a distinct, self-controlled formation. Moreover, Salmond brought to his post a near-unrivalled record of practical experience of air control, combined with swift intelligence, foresight, and a passionate devotion to his Service. Salmond was to remain CAS until 1 April 1933 on which date he was succeeded in the post by his brother Air Marshal Sir Geoffrey Salmond. During those three years John Salmond had to weather such obstacles as disarmament conferences, national financial crises, and revived attempts to – yet again – emasculate the RAF by reducing it drastically to a mere metropolitan aerial defence force comprised *only* of fighter aircraft; this latter idea was mooted at the 1932 Geneva Conference on disarmament and supported by the British Prime Minister, Ramsay MacDonald, Anthony Eden, and Sir John Simon. Fortunately for the RAF, and Britain, the Geneva Conference was ineffective and the British government, albeit reluctantly, finally reversed its policy and set about expanding and re-arming its Services. In the midst of these many problems, John Salmond quietly commenced a variety of innovations which would ultimately improve and benefit the RAF technically, in particular in the matter of flying training, aero-engine development, and re-equipment of his firstline squadrons with post-1918-designed aircraft. It could be truly stated that whereas Hugh Trenchard had laid the permanent foundation stones of the RAF, John Salmond then erected the overall superstructure for the future. His brother Geoffrey who succeeded him as CAS was fated to die within weeks of his appointment, on 27 April, and on 22 May 1933 the next CAS, Air Chief Marshal Sir Edward Ellington, took up the supreme RAF appointment.

When John Salmond became CAS in January 1930 he inherited an RAF comprised of a total of 57 squadrons – 22 of these based overseas – plus nine AAF or Special Reserve squadrons of the 'Weekend Air Force'. Four years later, on 1 January 1934, RAF strength consisted of 3,334 officers, 83 cadets, 25,170 non-commissioned airmen, and 1,794 aircraft

MRAF Sir John Salmond, GCB,CMG,CVO,DSO who succeeded Trenchard as CAS on 1 January 1930

apprentices under tuition. Squadron strength had grown to some 90 squadrons (including the Fleet Air Arm), and in March that year the Air Estimate amounted to £20,165,600, of which slightly more than one million pounds was the FAA's appropriation, an increase on the previous year's Estimates of a mere £135,000 net. Only in the following year 1935 did this RAF budget begin to show any intention on the government's part to actually commence real expansion of the long-neglected air service, when the 1935-36 Air Estimates showed an increased of £3,089,000 over its preceding figure.

The need for expansion and, especially, modernisation, was evident on 6 July 1935 when HM King George V carried out a Royal Review of the RAF at the part-completed RAF station at Mildenhall. A total of 356 aircraft were 'on parade' from 37 squadrons – and every one was a biplane, ranging in vintage from five to fifteen years in design. Just four months after that review, however, on 6 November 1935, a new monoplane fighter, serialled K5083, made its first testing flight – the prototype of the Hawker Hurricane; while on 5 March 1936 came the first flight of the prototype Supermarine Spitfire. Both designs exemplified the concern of British private aircraft manufacturers about the RAF's equipment, as opposed to the apparent complacency and lack of interest over previous decades by governmental departments.

The years 1935-38 might readily be termed the 'Panic Period' in the context of RAF expansion. Finally convinced of the futility of the many preceding disarmament conferences, the government hastily attempted to make up for its many years of neglect of the air arm by successive and rapid expansion schemes, aimed primarily at placing the RAF numerically (at least) on a par with the air forces of France and Hitler's Nazi Germany. Even so, priority was still given to Home Defence air power – the relatively few squadrons based abroad trying to maintain 'peace' in Britain's vast global empire were to remain last on the list for strengthening or modernisation in aircraft. In India, for example, Westland Wapiti aircraft first introduced to the sub-continent in late 1928 were to remain squadron equipment with certain units until 1940; a design which itself was simply a development of the 1918 DH9A in most respects ...

The pace of the hoped-for expansion of the RAF may be judged in part by a comparison of Air Estimates during the three years 1936, 1937, and 1938, these being £39 millions, £88½ millions, and £111½ millions respectively, apart from separate appropriations from the Navy Estimates to finance the Fleet Air Arm. Nevertheless, the mere provision of cash could not offer any *immediate* results in terms of modern aircraft and facilities, greater numbers of squadrons, more *trained* men, or updated strategic policies. The barren years of Trenchard's reign in the 1920s and vacillation of succeeding governments in the early 1930s were now clearly evident in default of a strong RAF. Evident too was the wisdom of Trenchard's prime policy of 'building a cottage on the foundations of a castle' – the RAF was at least now *capable* of being expanded on its existing structure.

Leaving aside the political and inter-Service polemics during 1918-38, what of the men who served in the RAF? In the early 1920s the RAF inherited both the best and the worst facets of its army and navy ancestors. Discipline was a curious bastard mixture of Kipling's 'Soldiers Three' and Nelsonian 'traditions', intermingled with regulations and practices created additionally specifically to encompass situations peculiar to RAF life. The more senior officers were, with rare exception, class-conscious and therefore unable easily to adjust to the fresher, more liberal attitudes of their juniors, thereby – albeit often quite unconsciously – creating a barrier between officer and ranker. Only on the firstline squadrons were such barriers well-diluted, where every man, of whatever rank or trade, quickly formed a tightly-knit, interdependent unit community, though without corrupting necessary overall discipline so vital for ultimate efficiency in any Service formation. The elementary code of discipline expected of all non-commissioned and junior commissioned ranks was initially inculcated at the 'Depot' – RAF Uxbridge – under the eagle eyes of such legendary characters as 'Stiffy' Wombwell; a time of recruit initiation well, if somewhat overdramatically described in T.E. Lawrence's *The Mint*.

For the non-commissioned pay at all levels was low, uniforms and equipment outdated, and living accommodation too often sub-standard or even near-primitive with regard for day-to-day facilities. The contemporary need for 'guarding the Empire' meant that any airman or officer serving on a regular basis was liable to be posted overseas for at least one full tour of duty; this tour being consolidated as a five years' stint from early 1924. This meant total separation from families for the vast majority of men, due to the lack of finance to provide married quarters on most overseas stations. In any case, junior officers were not granted a marriage allowance until they reached the rank of squadron leader or the age of thirty, whichever came first.* Thus, in the main, life on any firstline squadron overseas became primarily a bachelor existence, with little off-duty form of recreation available beyond that which the men created for themselves.

Under such conditions life on any overseas unit became almost wholly focussed on the aircraft and work, yet with the traditional perverseness of the British 'regular' Serviceman such circumstances merely served to intensify his natural determination to enjoy himself whatever the situation. Thus a spirit of true comradeship was created which has never been surpassed in the RAF, although equalled in certain cases during 1939-45 operations. The tone of that spirit is accurately reflected in the Service songs born in those years. With a traditional flavour of cynicism, fatalism, and twisted humour, these became part of RAF folklore; solace from stark surroundings and often grim conditions. Death and injury were no strangers to men flying over the towering mountains of India's North-West Frontier or the baking plains and deserts of Iraq. Crewing aircraft of ageing design, backed by minimal maintenance facilities, in climates and conditions never envisaged by the machines' designers years before, all air crews were well aware of the hazards involved but accepted these as routine facets of their duties. And it should be noted that parachutes for flying personnel were not a mandatory item of equipment until 1928-29.

* To attain squadron leader rank before age 30 at that time was virtually a pipe-dream for any junior officer.

On 1 September 1937 Sir Edward Ellington was succeeded as CAS by Air Chief Marshal Sir Cyril Newall. By then relatively great changes were afoot in the context of expansion and modernisation of the RAF, though the most significant of these, e.g. modern heavy monoplane bombers, were still in embryo stage only and would need several years to mature to actual operational use. Only in Fighter Command was there any material advance with the gradual introduction of Hurricane and Spitfire fighters during 1937 and 1938 respectively. Scheme after scheme was propounded for expansion of the RAF in the contemporary political and Service climate of anxiety about achieving parity (numerically) with Nazi Germany's Luftwaffe, and the latter's presumed capability of delivering a devastating knock-out blow at the commencement of any war with Britian.

These fears, exacerbated by overdramatically pessimistic pronouncements by the Joint Planning Committee of the Chiefs of Staff, and no less by Winston Churchill's over-exaggerated claims for the Luftwaffe's existing strength in aircraft, at least had the unsought merit of loosening purse-strings at the Treasury despite the latest Prime Minister, Neville Chamberlain's specific antipathy towards generosity in any Service financing.

For Cyril Newall his first year in the driving seat of the RAF was to be the most difficult year of any CAS to date. While fully occupied in speedy re-armament of the RAF, Newall's tenure coincided with Hitler's occupation of Austria and, later, Czechoslovakia, the Spanish civil war, and Japan's war with China in the Far East, apart from hostile anti-British propaganda from the Italian dictator Mussolini in the Mediterranean zones of British influence.

If all these matters assumed a first priority in the minds of Service and political chiefs in Whitehall, they little affected the RAF commands overseas as yet. Internal disturbances in Palestine and neighbouring Arab countries continued to occupy fully the few RAF squadrons available there, while on the North-West Frontier of India (now Pakistan) the pace and colour of daily RAF operations were much as before and, indeed, were to continue to be until long after war with Hitler's Third Reich erupted in late 1939. It should be remembered that the RAF in the Middle East did not commence true war operations (with Italy) until the summer of 1940, while in the Far East – to use a generic term for India, Malaya and Hong Kong *et al* – war with Japan only commenced at the end of 1941. Thus the so-termed Munich Crisis of September 1938 had no material repercussions on those RAF units based on outposts of the British Empire then in the short term. They were, however, to suffer neglect and long overdue replenishment and expansion when war finally came to their doorsteps as a direct consequence of the desperate struggles in Europe in 1939-40.

By the time of the Munich Crisis of September 1938 expansion of the RAF was well under way but as yet ineffective in terms of the force's updated operational capacity, had Britain been forced to declare war with Germany at that moment. As already stated, this state of RAF affairs in the UK had little bearing upon the tasks of those RAF squadrons based overseas beyond placing them last on the ladder of priorities – a status they were already well used to over the previous two decades. In September 1938 a total of 28 RAF squadrons were based overseas, as under:

Squadron No	*Aircraft*	*Main Base*
Middle East		
6	Hawker Hardy	Ramleh
8	Vickers Vincent	Khormaksar
14	Vickers Wellesley	Amman
30	Bristol Blenheim I	Habbaniya
33	Gloster Gladiator	Heliopolis
45	Vickers Wellesley	Helwan
47	Fairey Gordon/Vickers Vincent	Khartoum
55	Vickers Vincent	Habbaniya
70	Vickers Valentia	Habbaniya
84	Vickers Vincent	Shaibah
202	Saro London	Ras-el-Tin
203	Short Singapore	Basra
208	Hawker Audax	Mersa Matruh
211	Hawker Hind	Helwan

Squadron No	*Aircraft*	*Main Base*
216	Vickers Valentia	Heliopolis
223	Vickers Wellesley	Nairobi
India		
5	Westland Wapiti	Risalpur
11	Bristol Blenheim I	Risalpur
20	Hawker Audax	Peshawar
27	Westland Wapiti	Kohat
28	Hawker Audax	Ambala
31	Westland Wapiti	Drigh Road
39	Hawker Hart	Risalpur
60	Westland Wapiti	Kohat
Singapore		
36	Vickers Vildebeest	Seletar
100	Vickers Vildebeest	Seletar
205	Short Singapore	Seletar
230	Short Sunderland	Seletar

These 28 squadrons represented some 20 per cent of total RAF squadron strength at that period, and it will be readily noted that all but six of the above list were still flying biplane aircraft, while the sole pure fighter squadron was No 33, stationed in Egypt. The list does not include the many Fleet Air Arm squadrons and Flights in existence at that date. Although nominally still controlled by the RAF, in fact the near-twenty years' unceasing campaign by the Admiralty to gain full possession of its own FAA had virtually achieved its aim in the previous year when, on 30 July 1937, the Prime Minister, Neville Chamberlain, had announced his government's decision to instigate transfer of total control of the FAA to the Admiralty, though actual administrative control in the event was not to be assumed until 24 May 1939. At the same time it was made plain that RAF Coastal Command – which the Admiralty also coveted – would remain under the full aegis of the Air Ministry; a decision which would continue to be challenged by certain senior naval and political figures for several years to come.

Analysis of Britain's defences and political attitudes at the time of the 1938 Munich Crisis has been made in a plethora of published works over the past 50 years – much of this with the golden advantage of hindsight and latter-day moralisation – yet one facet was abundantly obvious at that time: the outdated and seriously understrength RAF in spite of many very recent hasty moves to rectify the situation. Yet had it not been for the unceasing endeavours of such visionaries as Hugh Trenchard, John Salmond and a few contemporaries there might not have existed *any* air force worthy of the description in 1938. The Shoe-string Air Force of the 1920s and early 1930s, so fiercely defended and gently nurtured by Trenchard and Salmond, and no less by their subordinates right down through the ranks of the RAF, was at least of high quality in men and standards. It now needed merely time and re-equipment to raise it in strength and status to its rightful position as a prime defender of Britain and the true vehicle of any future offensive capability.

CHAPTER TWO

Russian Interlude

In view of the past 40 years of East-West political tensions, it is perhaps occasionally forgotten that during most of the 1914-18 war the Imperial Russia of the Czars was Britain's ally in the struggle against the Central Powers – indeed, Russia was formally at war with Germany from 1 August 1914, three days before Britain declared war with that country, and on 5 September 1914 joined with Britain and France in signing the so-termed Pact of London whereby each signatory agreed not to seek any separate peace terms with any of the Central Powers.

Even less publicised is the fact that from 1918-20 the British Services – army, navy and RAF – not only fought alongside Russian forces but also *against* other Russians. Such a circumstance had resulted directly from events within Russia which came to a head in 1917, when in March a massive internal revolution commenced, resulting in a new Provisional Government being proclaimed on 14 March, followed next day by the abdication of Czar Nicholas II. Before the end of March Britain along with her allies France, Italy, Rumania and the USA had officially recognised the new Russian Provisional Government which, on 15 September 1917, declared Russia to be a Republic.

A *coup d'état* in early November 1917 gave Lenin and Trotski power with their Bolshevik movement and led to initial feelers for an armistice between the Bolsheviki and the Central Powers on 21 November. By December 1917 hostilities between the Russian and German armies had begun to be suspended, and on 15 December an armistice was officially confirmed between the Bolshevik Government and the Central Powers at

Brest-Litovsk, this to commence from noon on 17 December and terminate on 14 January 1918.

The Brest-Litovsk negotiations proved lengthy and over the next few months Russo-German conflict resumed briefly before a final cessation of the war was announced by the Bolshevik authorities. This Bolshevik revolution was by no means wholly supported by all the Russian people from the outset, and even as the Trotski/Lenin government debated peace terms with the Central Powers anti-Bolshevik Russian armed factions were in conflict with Bolshevik forces in various regions. Space does not permit even a reasonable summary of the subsequent internal wars and conflicts which followed during the bloody years 1918-20, these being highly complicated and involving a bewildering variety of Russian and non-Russian armies and other forces. Some of those forces, however, were British, including small elements of the recently formed RAF.

The signing of the final peace treaty between the Bolsheviks and the Central Powers in early 1918 gave Britain serious cause for anxiety. The cessation of hostilities on that eastern front would (and did) release massive German army forces for reinforcement of the existing German armies along the Western Front opposing British and Allied forces. In addition Germany might now establish submarine bases at such ports as Archangel and Murmansk and use these for raiding Allied shipping in the North Sea; while the rich resources of western Siberia would now become available to the Central Powers.

These, and other no less important, considerations led the British immediately to despatch a mixed expeditionary force to Murmansk, including a Flight of De Havilland 4 bombers aboard HMS *Elope*, which were joined in July by a complement of five Fairey Campania and two Sopwith Baby floatplanes and a single Sopwith Camel aboard the carrier *Nairana*. These RAF units took part in the occupation of Archangel on 2 August 1918, and in the following month were used to cover Allied infantry – by bombing and strafing attacks – when these successfully occupied Obozerskaya, some 100 miles south of Archangel. In October 1918 the *Nairana* returned to England,

Loading a Short 184 seaplane, N9290, on *HMS Pegasus*, bound for Russia, 1919.

RE8 of the RAF 'Detachment' in North Russia, 1919

and by December the small RAF contingent remaining in Russia had been organised into two separate commands; one at Archangel and the other based at Murmansk.

With the defeat of Turkey and the Central Powers by November 1918, Britain and her allies despatched various forces to 'protect' the territories surrendered by the Turks and Germans in the Caucasus, and these forces included No 221 Squadron, equipped with a mixture of DH4s, DH9s and DH9As for bombing, with a few Sopwith Camels as escorts. 221 Squadron had been formed from 'D' Squadron, 62nd Wing RNAS based at Stavros in 1918, and by December that year, commanded by Major J.O. Andrews, DSO,MC,* was ordered to move to Batum at the eastern corner of the Black Sea. Embarked on the seaplane carrier HMS *Riviera*, 'A' Flight and the squadron's HQ staff sailed on 29 December, reaching Batum, via Constantinople, on 3 January 1919. From here the unit made a train journey to Petrovsk, a port on the Caspian north of Baku (now titled Makhachkala), arriving on 12 January. All aircraft and equipment were quickly unloaded and prepared, despite conditions described graphically in the diary of D.B. Knock, a squadron armourer:

> *14/1/19.* Weather atrocious. Below zero. Wind screens erected and riggers and working parties have terrible job assembling Nines [DH9s] and Nine Acks [DH9As] in the open. Water has to be boiled and oil heated before filling radiators and sumps. A Nine takes off for test and returns into gale and blizzard. Pilot and AM [Air Mechanic] passenger almost frozen. No more flying until this wind drops.†

By the end of January most aircraft had been assembled and were ready for operations. Two further extracts from Knock's diary illustrate the somewhat confused situation existing in the area in the context of the various Russian warring factions:

* Later Air Vice-Marshal, CB,DSO,MC.

† *Leaves from an Armourer's Diary* by D.B. Knock: *Popular Flying*. May 1938

5/2/19. Realise that we are actually attached to Deniken's White Army here. Few months ago the Whites drove the Reds back north, where they hold Astrakhan.

10/2/19. Two Nine Acks go on reconnaissance north, each with two 65 and one 230lb bombs. South of Astrakhan they spot a parade of cavalry with red pennons flying. Drop the pills and get photos. Return and develop in glee showing much carnage. One up against the Reds.

13/2/19. Cossack officer of high rank arrives with interpreter. Story gets around. That was no Red cavalry bombed, but the side we are supposed to be assisting. Our 'planes had decimated a squadron of White cavalry. W/T man tells me that SNO (*Senior Naval Officer*), Caspian has ordered 221 to cease activity until further orders. *Thought* that Cossack looked furious. No wonder!.

This order to stop operations proved temporary and 221's DH9As were soon flying further bombing sorties against 'Red' Russian (i.e. Bolshevik) forces at Astrakhan and other locations. In March 1919 the unit was joined at Petrovsk by No 266 Squadron from Malta, equipped with Short 184 seaplanes. In the event, both squadrons were to remain on active operations until both were disbanded on 1 September 1919.

Knock's references to Whites and Reds reflected the common terms for anti-Bolshevik and Bolshevik (i.e. communist) forces respectively, and the British government had long decided to assist the White Russians in their struggles to defeat the revolutionary Reds. This direct operational assistance quickly raised much opposition to any such 'foreign entanglements' (*sic*) in the UK Parliament, resulting in the withdrawal of British forces from the Caucasus by late August 1919 and disbandment (officially) of Nos 221 and 266 Squadrons, leaving only a token guard force at the port of Batum.

In the interim, however, a strong military 'Training Mission' was assembled in Britain at the behest of the recently appointed Minister of War, Winston Churchill, who, from his first day in his new office, never ceased to work for the destruction of

Bolshevism. In order to placate the many critics this mission was designated an 'advisory force' intended merely to assist in training White Russian forces in the use of various arms being supplied to them from British sources. Part of this mission was an RAF contingent to be based at Taganrog, at the north-east point of the Sea of Azov, commanded by Lieutenant-Colonel A.C. 'Cissy' Maund, whose prime task was to train Russian air crews to fly a 130 RE8 aircraft given to the White Russians by Britain. This mission left Britain in April 1919 and arrived in southern Russia in the following month where it was to be supportive to the White Russian armies commanded by General Anton Deniken, the most successful of the many counter-revolutionary army commanders.

The fiction of the RAF mission being purely a training formation was revealed by the inclusion of an operational RAF unit intended to undertake actual war sorties on behalf of Deniken's armies against the revolutionaries. This unit, No 47 Squadron RAF, had been stationed in the Macedonian theatre of war since September 1916, operating as a mixed bomber unit against Bulgarian and other Central Powers' forces, and in April 1919 the squadron commenced a move to Novorossisk, a port on the Black Sea, from its former base at Salonika.

By the end of May the squadron was fully re-assembled with its headquarters at Ekaterinodar, some 60 miles north-east of Novorossisk, under the temporary command of Captain S.G. Frogley, RAF, equipped with a mixture of DH9 bombers and a few veteran 'hack' aircraft still lingering from the war. This move of 47 Squadron had been on a non-voluntary basis – a case of 'orders is orders' in Service vernacular – but in England the officially appointed squadron commander, Major Ray Collishaw, DSO,DSC,DFC, the noted Canadian fighter ace already credited with some 60 aerial combat victories in the 1914-18 war, was busy recruiting true volunteers to join his command and serve in south Russia against the 'Bolshies' (*sic*).

Touring around numerous stations in Britain, Collishaw found plentiful young air and ground crews ready to volunteer to join him, and in late May, with ten officers and 255 non-commissioned airmen, he left England, crossed France

Fairey IIIC, N9241 of the RAF 'Detachment', snapped at midnight in the summer of 1919 on River Dwina, Bereznik, North Russia.

North Russia scene, 1919. From rt, Sopwith Snipes E6360 & E6351, three DH9As, three Sopwith 1½ Strutters.

and Italy by train, then shipped the rest of the journey to Novorossisk, arriving there on 8 June 1919. Collishaw's appointment as official CO of 47 Squadron was dated as from 13 June, although he did not actually join his squadron at Ekaterinodar until 11 July after busy weeks of chasing up unit equipment and other stores. Before Collishaw's arrival on the squadron operations had commenced, when on 23 June three DH9s of 'C' Flight, detached to Gniloaksaiskaya, some 60 miles from Tsaritsin (later retitled Stalingrad), armed with sixteen 20-pound and two 112lb bombs, attacked the south-east rail station at Tsaritsin, bombing their target first then strafing barges on the Volga, nearby streets, and a gaggle of cavalry near Elshanka. It was the first of many daily raids against Red Russian forces occupying the city over the following days, assisting Deniken's armies in successfully forcing the Reds out of Tsaritsin by the start of July.

After Baron Wrangel's particular army had occupied Tsaritsin for the Whites, 'C' Flight moved forward to Beketovka, twelve miles south of the city, where with fresh crews sent by the squadron at Ekaterinodar it recommenced bombing operations on 23 July. Two days later 47 Squadron recorded its first aerial victory in south Russia when two DH9s attacked the rail station at Kamychin but on their return flight were in turn attacked by a Red Nieuport Scout fighter. One DH9 gunner, Lieutenant H.E. Simmons, MC, got a burst of 50 rounds into the Nieuport which flipped over sharply and crashed to earth.

On 30 July three DH9s set out to bomb and photo-recce Tcherni-Yar and provided an outstanding example of the courage of the squadron's crews. The three crews were Captain S.G. Frogley/Lieutenant N. Greenslade, Captain W.F. Anderson/Lieutenant J. Mitchell, and Captain W. Elliot/Lieutenant H.S. Laidlaw, and once bombing had been completed Anderson went lower to let his observer take photographs of the results, accompanied by Elliot's DH9 for protection against any possible air attack. Both DH9s ran into a hail of heavy groundfire, and Anderson's starboard fuel tank was hit and leaking petrol fast. His gunner, Mitchell,

(*Right*) Major Ray Collishaw, DSO,DSC,DFC, OC 47 Squadron 1919-20 in South Russia.

(*Below*) Captain S M Kinkead, DSO,DSC,DFC with his Sopwith F1 Camel of 'B' Flight, 47 Squadron, South Russia, 1919. Roundel colours were (from outer ring) red, blue, white.

immediately clambered out onto the wing and plugged the holes with his fingers.

As Anderson then turned to return to base he saw Elliot's DH9 go down to a forced landing, its engine shot out of action, near a large force of Red cavalry which promptly galloped towards the stricken bomber. Despite his own precarious circumstances, Anderson landed reasonably close to Elliot's machine and, after Elliot had succeeded in setting fire to his own DH9, picked up both Elliot and Laidlaw, then set off again under fire, with the gallant Mitchell remaining on the wing for the whole 110-miles trip, still plugging the petrol tank bullet holes and being steadily burned painfully by the aircraft's engine exhaust. The over-burdened bomber reached base safely, and Anderson and Mitchell were each subsequently awarded a DSO, and DFC in respect of later services in the Russian campaign. Elliot was later to complete a distinguished RAF career as Air Chief Marshal Sir William, GCVO, KCB,KBE,DFC.

The crews of 'C' Flight maintained their pace of operations in the month of August, bombing a variety of objectives successfully and in one case creating a huge fire at one Red airfield. An example of the intensity of the Flight's effort was the work accomplished on a single day, 5 August, when three DH9s each flew three sorties throughout the day against Tcherni-Yar, dropping a total of 67 bombs and firing 2,300 rounds of ammunition during sixteen hours' flying time. In August, No 47 Squadron began receiving some DH9A aircraft, ex-221 Squadron machines transported by the carrier *Ark Royal*, and some reinforcement personnel from various sources. These DH9As were sent to re-equip 'A' Flight, whose DH9s were passed to 'C' Flight. Then, in September, 'B' Flight was equipped with Sopwith Camel single-seat fighters, transferred from the ex-RNAS base at Mudros. These were quickly assembled and 'B' Flight, under the command of Captain 'Sam' Kinkead*, joined its companions at Beketovka by the end of the month. On 30 September Kinkead, while

* Later Flight Lieutenant S.M. Kinkead, DSO,DSC,DFC

escorting some DH9s on a bombing sortie, was attacked by a Red Nieuport fighter which he quickly shot down into the Volga.

In October 1919 No 47 Squadron *per se* ceased to exist officially with effect from 1 October, but remained as an operational unit now retitled 'A' Squadron. This paper disbandment had been the result of strong representations in Britain by various factions, including the embryo British Communist Party, for withdrawal of all British regular forces from Russia. Henceforth, 'A' Squadron was to be officially classified as a facet of the RAF Training Mission, but all members wishing to remain with the unit were required to sign a form *volunteering* to continue in support of General Deniken's White Russian anti-Bolshevik cause. In practice the former 47 Squadron simply signed the form and carried on its intense campaign of aiding the armies of Deniken and Wrangel.

In October the Red armies launched an offensive designed to re-occupy Tsaritsin from the north, supported by armoured trains, Red aircraft, and a flotilla of vessels on the Volga. The Red cavalry, some 5,000 horsemen led by their commander Dumenko, broke through the junction of the White Caucasian and Don armies on the river Don and approached Tsaritsin from the south-west, within 40 miles of his objective, when this force was found by the Camels of 'B' Flight, 'A' Squadron. Four Camels, led by Kinkead, caught the cavalry force on flat open ground and proceeded to slaughter it, keeping up deadly low-level strafing attacks which totally scattered the horsemen in all directions, creating heavy casualties. The fleeing cavalry was then cut to pieces by White Cossacks who later claimed to have found at least 1,600 dead on the killing ground.

Dumenko's attempted offensive had been effectively halted, and his division was not seen at the fighting front for some time thereafter. While Kinkead's Camel quartet was dealing with Dumenko's cavalry, 'A' and 'C' Flights were also engaged, attacking the Red fleet in the Caspian which was shelling Tsaritsin. Led respectively by Captains L.H. Slatter, DSC,DFC*

* Later, Air Marshal Sir Leonard, KBE,CB,DSC,DFC.

and S.G. Frogley, these two Flights pressed home their bombing attacks through heavy land-based and shipborne shell-fire and the Red fleet retired back up the Volga in disorder, its crews stating that it was 'murder to come within reach of the English aeroplanes' (*sic*).

By mid-October General Deniken's armies had reached a peak of successes against the Bolsheviks and were preparing to advance on Moscow, but then began suffering a series of reverses and were forced to retreat across former hardwon territory southward. At that period Ray Collishaw fell victim to typhus – a disease rampant throughout the area – and only eventually rejoined his unit on 27 November. Meantime 'A' Squadron continued its support of the retreating White armies, with its 'C' Flight now re-equipped with RE8 aircraft; its DH9s having been given to the White Russian aviation units. As Collishaw resumed his command the squadron was moved westwards to the Kharkov front to try helping to stem the Red advancing armies there, leaving only 'C' Flight behind on the Volga zone.

By 11 December 'A' and 'B' Flights were withdrawn to Taganrog, the HQ of the RAF Training Mission, to refurbish its aircraft and other equipment, but suffered the loss of 'B' Flight's aircraft and stores which had to be abandoned in view of the rapidity of the Bolshevik advance towards Kharkov. The Flight personnel managed to evade capture and rejoined the squadron at Ekaterinodar, where it was joined by 'C' Flight from the lost Volga front in late January 1920.

In March 1920 'A' Squadron received orders to evacuate south Russia from Novorossisk, and step by step, still operating against the advancing Red forces, the squadron made its way to the evacuation point. Retreating with Collishaw's squadron from the Kharkov front was another RAF unit, 'Z' Flight, commanded by Squadron Leader Joe Archer, which had been formed as a small separate unit from volunteers serving with the RAF Training Mission, equipped with RE8s, and sent to the Kharkov fighting zone by the beginning of December 1919.

In that same month 'Z' Flight was parked on a railway line north-east of Kharkov, with two RE8s ready for flight and four others part-dissembled on the flat cars of its private train, waiting official permission to proceed with a bombing raid against Moscow. Permission was curtly refused, and Archer was left with no alternative but to join the general evacuation of British forces. The journeys towards the evacuation port of Novorossisk became nightmares for all concerned, with the air crews continuing to attack their Bolshevik pursuers whenever an opportunity presented itself, but theirs was now a case of self-defence rather than offensive, planned operations.

Collishaw was among the men who flew to the very last, and in late February one of his sorties almost ended in disaster when his DH9A's engine was hit by groundfire and he was forced to land. Once on the ground Collishaw managed to obtain some power from the damaged engine and proceeded to taxi it nearly twenty miles to Djankoi and safety.

The last air attacks by 'A' Squadron were flown on 28 March 1920, bombing and strafing cavalry and an armoured train, and the ultimate flight next day by Collishaw, a sweeping reconnaissance only over Red territory. At Novorossisk the squadron's Camel aircraft were crushed to rubble by some army tanks, while other aircraft and supplies were handed over to a White Russian squadron on 30 March, after which the remaining personnel went to a port at Theodosia from where they were embarked on a British transport ship and sailed, via Constantinople, to England. In one year of almost non-stop operations over southern Russia the men of No 47/'A' Squadron(s) were awarded totals of four DSOs, two OBEs, two MBEs, sixteen DFCs and one Bar to the DFC, three AFCs, thirteen Meritorious Service Medals (MSMs) and a flock of 75 Russian orders and decorations of various sorts. The original unit title of No 47 Squadron had already been reborn before Collishaw's men left Russian soil, when on 1 February 1920 at Helwan, Egypt, No 206 Squadron, a DH9 bomber unit, was re-numbered as the new 47. Collishaw's own summation of his unit's efforts was:

> I feel that No 47 Squadron made a surprising contribution to the White cause, mostly through its attacks on the Red ground forces. The destruction that we were able to wreak

> on the enemy on numerous occasions boosted tremendously the morale of the White troops in the area and at times caused near-panic in the ranks of the Reds. Unfortunately, the primitive organisation of Wrangel's force, in support of which most of our operational flying was done, prevented our successes being fully exploited.*

While No 47 Squadron had arrived in south Russia in the early summer of 1919, a separate British force had been assembled and despatched to northern Russia to supplement the small British and Allied presence around the vital ports of Murmansk and Archangel. This 'Relief Force' (*sic*) included a relatively large RAF contingent. Its air crews had been almost wholly recruited from serving volunteers from the RAF – some as a result of publicly displayed posters calling for men to join this force, while many had 'enlisted' spontaneously at the behest of fellow officers. A high proportion of the RAF crews were ex-wartime veteran fliers, much bemedalled in some instances†, who openly sought any form of excitement that might relieve the tedium of peacetime Service routine; a perfectly natural reaction for young men who had left a school desk to fly and fight in the 1914-18 war and therefore knew no other sort of existence.

Typical of the old boy network form of recruitment was J.I.T. Jones, DSO,MC,DFC,MM, a fiery little Welsh pilot who had been credited with 40 combat victories in 1918 while serving in France with No 74 Squadron. Spotting a recruiting poster appealing for such volunteers, Jones went to the Royal Air Force Club in Bruton Street, London where he met another veteran fighter ace, Geoffrey H. Bowman, DSO,MC,DFC, a major in the postwar RAF.‡ 'Coming to Russia with me, Taffy?' was Bowman's greeting. Jones shook his hand and the posting was fixed.

This Relief Force began being shipped to Russia in early

* *Air Command* by AVM R. Collishaw/R.V. Dodds; Wm Kimber, 1973

† Among the mixed army/RAF personnel were several holders of the Victoria Cross, including Lieutenant Alan Jerrard, VC,RAF.

‡ Present RAF rank titles were not introduced until an official promulgation dated 27 August 1919.

June 1919 and disembarked personnel and supplies variously at Murmansk and Archangel; the RAF at the latter port then setting up shop on an airfield at Bereznik, some 140 miles from Archangel itself. The initial aircraft issued to the RAF were DH4s, DH9s and DH9As and a few Sopwith Camels and Snipes, while No 1 Slavo-British Squadron which shared the aerodrome (a polite title for what was merely an uneven surface grass space of very limited size) was equipped with a motley mixture of old Sopwith 1½ Strutter two-seaters (mostly rebuilt from crates found in the area left over from World War One), and various British DH9s and single-seat Sopwith fighters; this White Russian unit being commanded by Staff Captain Alexander Kazakov, the leading Russian fighter ace of the recent war. The British units became titled Nos 2 and 3 Squadrons, a purely local nomenclature having no connection with the RAF's regular units of the same numbers. No 2 Squadron (DH9s, DH9As and Sopwith Snipes) was commanded by a Major Carr, while No 3 Squadron (DH9s and DH9As) was commanded by Major Bowman. By July all units were in position to commence active operations against Bolshevik forces occupying the Dvina and Vaga fronts. The Dvina front was about 50 miles from the airfield while the Vaga area was only some fifteen miles away, both fronts being titled so from the rivers of the same name which conjoined locally. The RAF crews' main functions were to be bombing and strafing 'Bolo' forces – 'Bolo' being the common nickname given to the Bolsheviks by the British – although there were known to be Bolshevik air units based around the fighting zone equipped with ex-British, French and German aircraft of many types.

Once operations had begun the RAF crews found little air opposition but were constantly to run a gauntlet of heavy, accurate ground fire in their low level strafing-bombing sorties. Inevitably, casualties occurred but the most feared result of any such damage was the prospect of having to forceland in Bolo territory and be taken prisoner. In every area of fighting through the many Russian campaigns treatment by Bolsheviks of any captured opponent, irrespective of nationality or even sex, was notorious for its cruelty, even bestiality. Indeed, all

British Servicemen were advised to carry a personal sidearm to shoot themselves with rather than submit to any Bolo capture.

An example of such treatment occurred on 11 August when seven DH bombers from No 2 Squadron set out to attack the village of Gorodok. The bombers ran into thick fog, lost all sense of direction, and only two managed to return to base. One of the missing aircraft was crewed by Lieutenants D'Arcy Levy and Bugg which landed in unknown territory to discover their position. D'Arcy Levy climbed out of his cockpit to make enquiries of a soldier nearby, then realised he was in an enemy-occupied area. Dashing back, he climbed into his cockpit, tried to start the engine but choked it, then climbed out again to swing the propeller. Bolshevik soldiers ran to the aircraft and immediately bayoneted D'Arcy Levy to death, then dragged Bugg out of the machine. Fortunately for Bugg he was wearing a mechanic's pair of overalls over his uniform and was thought to be one by his captors, who spared his life and sent him back to prison.

The work of the Relief Force was destined to be relatively brief and by the end of August 1919 rumours – soon confirmed – were circulating of the imminent evacuation of the force from Russia. The crews continued flying against the Bolsheviks until the first week of September, but then received specific instructions to evacuate the airfield and withdraw to Archangel for embarkation aboard ships bound for England. For the majority of the RAF crews the whole episode had been something of a fiasco, and the general feeling was that the Bolsheviks should have been left to fight among themselves.

Their only regrets were in respect of the White Russians they had known and fought alongside, most of whom sadly rejected pleas from their British friends to accompany them to safety in England. The first rumours of the British evacuation, in early August, had depressed many White Russians, including Alexander Kazakov, commander of No 1 Slavo-British Squadron at Bereznik. On 3 August Kazakov ordered out his Sopwith fighter, climbed in, took off sharply, then from a height of merely twenty feet pulled the Sopwith up in a looping manouevre, stalled off the top and dived vertically into the

Funeral procession of Captain A Kazakov, 4 August 1919

DH9A of No.3 Squadron, RAF North Russia, with Captain J I T Jones in pilot's cockpit. Note 112lb RL bomb in foreground.

earth at full speed. Kazakov was buried next day in the nearby village with full military honours – a symbol of the hopeless White Russian cause.

The overall Russian intervention in the years 1918-20 had been a costly failure in many senses. Millions of pounds-worth of arms and supplies had been sent to the White Russians but had rarely been utilised effectively – much of the domestic stores finding its way immediately into the corrupt hands of profiteers and others – while thousands of tons of unused war material had often been abandoned by White Russian forces in retreat which served to equip the Bolsheviks. Even *before* the final evacuations of British forces from south Russia in 1920 the British government had begun trade negotiations with the existing Bolshevik 'government'.

Winston Churchill at the War Office had finally, albeit reluctantly, abandoned his hopes for crushing Bolshevism – a policy he had relentlessly pursued throughout despite a popular press campaign damning him for his views, and, inevitably, linking this latest Churchillian episode with his disastrous venture in the Dardanelles at Gallipoli during the recent war.

CHAPTER THREE

'Z' Force and the Mad Mullah

In the immediate post-1918 period Hugh Trenchard, struggling against military, naval and political opposition in his determination to establish the newly-created Royal Air Force as a permanent, independent Service, welcomed gratefully any opportunity or situation which would enhance his constant assertion that air power would prove more efficient and cost-effective than hitherto traditional, expensive army methods of settling disputes and rebellions in the many British overseas imperial and mandated territories. Fortuitously, a brief example of his contention occurred in 1919-20 which gave him a small but significant foundation block for his assertion. It happened in British Somaliland which had originally been a Dependency of Aden, then became a British Protectorate by treaty in 1898 when adminstrative responsibility for the area was transferred to the aegis of the Foreign Office. Its boundaries were agreed by Italy, France and Abyssinia, all of whom governed nearby territories bordering the new Protectorate. In 1905 overall responsibility was again transferred, this time to the Colonial Office.

Within the Protectorate the chief opponent to British 'protection' was Mohammed bin Abdullah Hassan, the self-styled 'Mullah of Somaliland' – or 'Mad Mullah' as he came to be titled by the British forces. Regarded as almost saintly by his highly superstitious followers, the Mullah had first achieved prominence in 1895 when he led a widespread rebellion against British administration, and by 1900 represented a serious threat to law and order within Somaliland. As a result punitive operations, involving much fighting and some 100 casualties, were undertaken by British army units in 1900-1901. The Mullah's forces withdrew eventually to beyond the Protectorate border where the British force commander, Colonel Swayne, was unable to puruse him. Further army operations in 1903, in conjunction with Italy and Abyssinia, cost the army ten officers and some 200 other ranks killed or maimed, with no decisive results. The resident army force was next reinforced to a total strength of 7,000 troops and placed under the command of Major-General Sir Charles Egerton, and ultimately a pitched battle was fought with the Mullah's forces at Jidballi on 10 January 1904, when the rebels were routed with a loss of over 1,000 men. The British losses amounted to 58 killed or wounded. Yet again, the Mullah escaped capture and fled across the border for safe refuge.

By 1909, however, the Mullah returned, instigating raids on pro-British tribes and generally spreading unrest. Faced with the prospect of yet more costly army punitive operations in a country larger than England and Wales combined, where the sole ground communications were restricted to camel tracks, the government decided to withdraw from the turbulent interior zones and to adopt a policy of coastal concentration. This rather negative policy soon proved unsatisfactory when the Mullah's men inflicted serious casualties on various outlying army formations, and in 1914 the administration decided once more to mount punitive measures. Before such expeditions could begin in earnest the outbreak of war with Germany intervened, resulting in the bulk of British forces in the Protectorate being withdrawn for service in other operational zones in the Mediterranean area.

Throughout the 1914-18 war the Mullah seized his opportunity to strengthen his grip on Somaliland, erecting stone fortresses which dominated the few passable roads into the hills, and his armed gangs soon controlled half the country and frequently raided and looted tribal villages in the other half, gaining a domination unable to be challenged effectively by the remaining weak and ill-equipped police and other government security forces. By 1919 the Mullah's powerful hold over Somaliland was so strong that Lord Milner, chief at the Colonial Office, could only advise the Prime Minister, Lloyd George, of just two alternatives to the situation. Either

Britain abandoned Somaliland to the Mullah's mercy, or it mounted yet another army punitive expedition on the pre-war model but of increased strength. This latter alternative, in the contemporary mood in Britain for strict economy in military matters, was financially prohibitive – the War Office admitted that at least two divisions would be needed and the 'campaign' would probably take a year to reach any conclusion – and therefore politically unpopular. As Milner's Colonial Under-Secretary, Leopold Amery, pointed out, millions of pounds and many lives had been spent pre-1914 in what he described as '... inconclusive, and more than once, disastrous fighting'.

Eventually, Milner, though no friend of Trenchard or his RAF, displayed courage by asking Trenchard to suggest some means of resolving the Somaliland dilemma. Trenchard's reply was succinct, 'Why not leave the whole thing to us?', meaning the RAF. The outcome was a meeting on 2 June 1919 at which Winston Churchill, Trenchard, Amery and the CIGS, Sir Henry Wilson – the latter being a bitter opponent of the RAF and Trenchard – discussed Trenchard's idea, then, despite the expected opposition from the CIGS, approved it.

The kernel of Trenchard's case was that no extra troops need be allocated to the existing residential forces apart from his RAF contingent – an argument which completely resolved the government's prime worry about the cost of any expedition. Now, with the Colonial Office's blessing, Trenchard set in motion the formation of a special RAF force for the task. Secrecy was paramount, so the projected formation was titled 'Z Force', and due to the total lack of local resources in Somaliland, the force was to be entirely self-contained in all respects in the contexts of maintenance stores and other supplies. The actual RAF force was to be comprised of twelve De Havilland 9 two-seat bombers and a European personnel strength of 36 officers and 183 airmen. The RAF commander of 'Z' Force was to be Group Captain R. Gordon, CB,CMG,DSO, assisted by Wing Commanders F.W. Bowhill, CMG,DSO* and W.

* Later Air Chief Marshal Sir Frederick, KCB,CMG,DSO.

Tyrrell, DSO,MC,* this last officer being the principal Medical Officer. The DH9s and the bulk of stores were to be supplied from depots in England, while other supplies and a proportion of personnel were to join the force from resident RAF units in Egypt, the nearest overseas RAF centre to Somaliland able to make such provisions.

While the aircraft and their spares were being crated in England – marked simply 'Middle East' with the mystic symbol 'Z' on the outsides of the crates – Gordon, Tyrrell and an 'Air Construction Service' (*sic*) officer left England as an advance party in late October 1919 for Egypt, then moved on to Berbera in Somaliland on 21 November, where their first priority was to select suitable sites for airfields and emergency landing grounds, now accompanied by certain other members of the 'Z' Force HQ staff. They finally selected Berbera as the force's main base with an advanced base at Eil dur Elan, to be supported as second-line air strips by landing grounds at Burao and Eil Dab once these two had been levelled satisfactorily.

In England meanwhile the main party of 'Z' Force left Victoria Station in the early hours of 13 November 1919 and proceeded to Alexandria, Egypt via Boulogne and Marseilles. On arrival in Egypt it was joined by other personnel detailed from local RAF units and, after adding further essential stores to those brought from England, embarked in HMS *Ark Royal* at Alexandria on 21 December, together with its aircraft, transport, and some 800 tons of supplies. After calling in at Suez and Aden, the force eventually reached Berbera harbour on 30 December 1919. Unloading commenced immediately upon arrival and erection of the DH9s began on New Year's Day. By 8 January three aircraft had been assembled and tested in the air, and by 19 January a total of eight DH9s were ready. Two other DH9s, D3117 and H5551, had been previously modified at the Royal Aircraft Establishment, Farnborough to become air ambulances, whereby one 'patient' could be accommodated in the normal rear cockpit in an inclined seat, with a second 'patient' laid on a stretcher under a protective

* Later Air Vice-Marshal Sir William, KBE,DSO,MC,MB,BCh,DPH,KHS.

DH9, D3117 of 'Z' Force, converted as air ambulance.

DH9, D3117 of 'Z' Force, with a patient being installed on the fuselage. Note mandatory spine pads worn by attendant airmen.

hinged canopy on top of the fuselage immediately to the rear of the second cockpit. At the same time as erection of the aircraft was taking place at Berbera, camel convoys set out daily from Berbera to convey supplies and personnel to the advanced base at Eil dur Elan, this air strip being declared usable from 17 January, and occupied by the eight DH9s three days later. Operations were then scheduled to commence on 21 January.

The first raid, by six DH9s, was flown against the rebel forts at Jid Ali and Medishi. Direct hits were observed and indeed this operation very nearly ended the operations immediately because one bomb penetrated the main building at Medishi Fort and detonated in the very room occupied by the Mullah himself, killing an aide standing next to the rebel leader and singeing the Mullah's clothing! Apart from some desultory rifle fire there was no opposition, although one DH9 was forced to land at Las Khorai due to engine failure. During the next two days further bombing raids caused heavy casualties among the rebels and their livestock at these two forts, but on the 24th an air reconnaissance showed the area to be apparently deserted apart from a few stray livestock, and it was assumed that the rebels had retreated towards another stronghold at Tale. On that same day ground troops captured Baran where another landing strip was then prepared, while on 28 January they occupied Jid Ali fort. It was decided by then that the Somaliland Field Force – consisting of detachments of the King's African Rifles, Somaliland Camel Corps, and 101st Grenadiers, Indian Army – should play a major role. Accordingly, from 25-30 January the RAF was mainly tasked to co-operate closely with this force, primarily by maintaining contacts with the various detachments and message-dropping, and utilising a temporary air strip at El Afweina for these duties.

On 29 January a fort at Galibariboa was bombed together with its surrounding native huts, although few rebels were observed, but next day, resulting from information given by a surrendered sheikh that the Mullah and his men were making for their Tale stronghold, the OC Field Force decided to give immediate chase in co-operation with the RAF. The advanced air base was moved to El Afweina and on 31 January the Mullah's baggage column and personal following was spotted, then attacked by the patrolling DH9s. The Mullah once again had a narrow escape, hiding in a nearby *nullah* (valley).

Meanwhile the Camel Corps had continued its pursuit and arrived at El Afweina on 1 February, on which date aircraft reconnaissance of Tale indicated a large mass of livestock there which were promptly attacked. On 4 February Tale Fort was bombed by three DH9s and several directs obtained including one on the Mullah's private enclosure just outside the perimeter, and setting fire to many native huts with incendiary bombs. At that stage it was intended to get the ground forces to close on Tale after more bombing attacks, but the latter proved unnecessary because troops caught the Mullah's convoy and then captured Tale, while the Camel Corps intercepted and destroyed the rebel leader's personal party which had fled from Tale after the first bombing raid.

Once more, the Mullah himself evaded capture and fled, with only four other men, over the border into Abyssinian Somaliland. There he was to remain, his influence nullified, until his death in 1921 (one unconfirmed report stating that he was assassinated), at Imi.

All operations being clearly concluded, all aircraft returned to Berbera on 18 February. During the whole campaign, apart from cases of illness, no British casualties had been incurred – a complete contrast to all previous army campaigns against the Mullah – while the DH9 ambulance-converted aircraft had proved ideal for evacuating a total of eight men from Eil dur Elan to Berbera's hospital facilities in mere hours instead of the men having to endure lengthy journeys by road. With the operations completed well before any anticipated date, the RAF force found itself with some two months to while away before the scheduled re-embarkation for Egypt or England, and used this unexpected 'holiday' to clear up all administrative work, hand over any stores of possible use to the local Public Works Department, including most of the force's 21 vehicles eventually, and, after a few demonstration and 'joy-ride' flights around Berbera, dismantling the aircraft and crating same ready for the final move. Finally, 'Z' Force embarked on RIMS *Dufferin* in mid-April and sailed for Egypt where its personnel were then dispersed to various new postings.

The success of 'Z' Force, in co-operation with the Somaliland Field Force, was hailed in England as proof of Trenchard's prime advocacy of what would come to be termed 'air control' of overseas territories. The Colonial Under-Secretary, Amery, pointed out that, at a total cost – including army transport and pay, as well as the RAF – of £77,000, the Mad Mullah final campaign was ' ... the cheapest war in history'. Trenchard himself was jubilant about the result, which appeared to justify all he had been claiming for the Third Service.

Viewed pragmatically, the 'Z' Force operation was simply a one-off situation, however well conceived and executed, and not a wholly Air Force operation; a view seized upon by certain disgruntled army and other opponents of Trenchard in the continuing attempts to dismember the RAF in following years. Nevertheless, the Somaliland campaign had at least sown a seed in political minds, particularly Churchill's, when native rebellions became widespread in Mesopotamia in late 1920. He (Churchill) immediately consulted Trenchard to ask him to prepare a rough plan for policing Mesopotamia (later, retitled Iraq) by air.

CHAPTER FOUR

Mespot

With the fall of the Imperial German and Ottoman (Turkish) dynasties in late 1918, several former German and Turkish territories were occupied and taken over in the Middle East by the Allied victors and, under a system adopted by the League of Nations, these were created as new States. Since none of these was as yet capable, financially or politically, of true self-government, the League devised an interim measure whereby such states became mandated to various Allied governments who then assumed responsibilities for administration, internal and external security, and progressive development of their mandates towards eventual independence in their own right.

Under this scheme the United Kingdom inherited the problem states of Palestine, Jordan and Mesopotamia in early 1920; the latter – known as 'Mespot' to all British Servicemen – being a loosely defined area encompassing much of Persia (now Iran) and the present-day Iraq. This bland system of geographical redefinition by the League was immediately greeted with hostility from Turkey, which blankly refused to recognise any such new 'state', and was equally unacceptable to the premier Arab rulers of the time, Ibn Saud, Sheik of Nedj (later, King of Saudi Arabia). The result was a rapidly spreading wave of tribal clashes and anti-British uprisings throughout the British mandated territories, particularly in Mesopotamia.

These rebellions in 1920 had been preceded by a series of anti-British revolts engineered by one Sheikh Mahmud, head of the family of Barzanji Saiyids. Son of the corrupt and tyrannical Sheikh Said, the great-grandson of Sheikh Kaka Ahmad, a Kurdish holy man of 'fabulous virtue and power' (*sic*), Mahmud's influence over the Kurdish inhabitants of all districts south of the Greater Zab river was potent. His over-riding ambition was to create an independent Kurdistan under his personal sovereignty, and his first clash with British forces had been at the battle of Shaibah in April 1915 where he had been quickly defeated. Three years later Mahmud stirred up a minor revolution and, in May 1918, was appointed by the Turks as Governor of Sulaimania, pledged to oppose the British.

Despite this pledge Mahmud intrigued with the British who, in December 1918, made him head of a 'committee' (*Hukamdar*) to govern Kurdistan from the Greater Zab to the Diyalah. Characteristically, Mahmud merely used this position to increase his influence over the many tribes of Kurdistan – a nebulously defined area covering northern Mesopotamia, western Persia, parts of southern Russia and the former Ottoman empire.

In May 1919 Mahmud attempted a *coup d'êtat* by assuming mastery of Sulaimania and interning all British political officers resident there, but a month later his forces were resoundingly defeated at the Bazian Pass. Mahmud was captured, court-martialled and sentenced to death, but then had his sentence commuted to ten years' imprisonment and sent to India. In the event his 'banishment' proved to be relatively brief and Mahmud was destined to return to Iraq and continue his crusade over the following decade.

Throughout these uprisings in 1919-20 the RAF in Mesopotamia, Persia, Palestine and Jordon was pitifully small at first. In 1919 only two squadrons existed in these areas, these being Nos 6 and 14. No 6 Squadron, equipped with RE8 aircraft, moved from Spa, Belgium to Basra, via France and Egypt, arriving at Basra in mid-July 1919; while 14 Squadron, flying Bristol F2Bs, was based at Ramleh in Palestine. A third resident unit, No 30 Squadron, existed, albeit at near-cadre strength and much dispersed by Flights until being officially restored to full squadron status with effect from 1 February 1920. No 30 Squadron, based at Baghdad West, was equipped with a mixture of RE8s and DH9As, and despite its legal restoration to squadron establishment continued to have its Flights widely dispersed until finally coming together as a complete entity in Hinaidi in October 1927.

Nos 6 and 30 Squadrons, in spite of a total strength of little more than a dozen serviceable aircraft between them by Christmas 1919, were called upon to undertake spasmodic operations against unruly tribes. By May 1920 general insurrection throughout Mesopotamia and Persia was reaching boiling point, with constant raids on main lines of communication, besieging of various outlying British garrisons and other posts, and finally becoming a virtual all-out revolt by July. British military forces in Iraq at that time totalled slightly more than 120,000 troops, with a further 13,600 in Persia, but all were very thinly distributed in small garrisons across the vast lands. In August-September an additional 15,000 British and Indian troops were sent as reinforcements from India, though again these tended to be distributed piecemeal rather than concentrated as a fighting force of any genuine strength.

For the first three months of unceasing operations during June-August 1920 the resident RAF squadrons, Nos 6 and 30, soldiered on manfully without reinforcement or significant increases in either aircraft or personnel. In *shade* temperatures ranging betwen 110-118 degrees, where even an initial climb to 1,500 feet at dawn could use most of the water in engine radiators, the air crews had their hands full simply controlling their aircraft. Some indication of such conditions were expressed by Squadron Leader G.C. Pirie, MC,DFC, commander of 6 Squadron*, when referring to a particular operation to attack rebels besieging the garrison at Rumaithah:

> We would leave Baghdad at about 0400 hours and fly to Diwaniyah, consult there with the Political Officer and garrison commander, then proceed to Rumaithah. While one machine took messages off the Popham Panel in the serai, the remainder of the Flight bombed and machine-gunned any gatherings seen nearby.

* Later, Air Chief Marshal Sir George, KCB,KBE,MC, DFC, LLD.

Vickers Vimy of No.4 Flying Training School (FTS), Abu Sueir, Egypt. No. 4 FTS was the prime training school in the Middle East throughout 1918-38.

DH9As of 'B' Flt, 55 Squadron at Mosul, ca, 1921. 'B3' is E9911 & 'B4' is E9912.

Vickers Vernons of 45 Squadron at Abu Sueir, 7 February 1922.

After replies had been dropped to the Panel messages, the Flight would proceed to Samawah, where the garrison was encamped on the aerodrome. Filling up our machines there was a laborious and unpleasant task, for we had to sit on the top planes of the Bristol Fighters and fill the tanks from five-gallon tins. The fierce hot wind from the north blew much of the petrol on to our clothes; these, consisting as they did of a pair of shorts and a shirt, were no protection against the scorching and stinging pain of evaporating petrol.

After a hasty lunch, which we had brought with us, we used to load up with bombs and proceed to attack the tribes west of Samawah. This accomplished, we returned to Samawah, fixed on more bombs and started on the homeward journey. Practically every afternoon at this time a tremendous sandstorm began to blow about 1400 hours. Through this we had to fly – no simple matter. Even at 8,000 feet the sand swirled about the machines, and only by straining one's eyes to the utmost could the winding course of the Euphrates be followed. It was almost impossible to control an aeroplane under these conditions, and when we descended over Rumaithah to see that the garrison were all right, and to bomb gatherings nearby, we felt more anxious about the ability of the machines to stand the buffeting of the storm. Landing at Hillah on the homeward journey to report, we usually arrived back at Baghdad at about 2000 hours – long and trying days, but inevitably so, in view of the shortage of machines and pilots.*

Throughout July-August the *average* availability of aircraft was merely six Bristol F2Bs and five RE8s on 6 Squadron, with three DH9As and two RE8s from 30 Squadron, although the RE8s were seldom used and the DH9As fully employed keeping communications open with Samawah. Operations were to continue apace until the end of 1920 by when two more squadrons entered the arena. No 84 Squadron was reformed at Baghdad with DH9As in August 1920 and commenced

* Lecture to Staff College, 1924.

Bristol F2Bs (nearest J6648) of 14 Sqn at the official opening of Kalundia aerodrome, Jerusalem, May 1924.

DH9A E8650 of 'A' Flt, 84 Squadron, Shaibah, 1923. Pilot here was Flg Off F F Inglis (later, AVM,CB,CBE). Note playing card symbols, an 84 Sqn traditional marking for many decades; spare wheel *under* fuselage; tropical radiator; and bomb-carrier under fuselage.

Bristol F2Bs of No.6 Squadron undergoing AOC's inspection by AVM Sir Robert Brooke-Popham, KCB, 1928. Note all propellers are *exactly* horizontal – checked by plumb-line!

DH9A, J7124 of 30 Squadron at 1000 ft over the Bazrian Ridge, Iraq on 6 August 1924.
Note all wing-tips painted red for easy identification if force-landed.

DH9A, F2858, 'M' of 8 Squadron, 1924 undergoing an engine change.

DH9A, E944, 'X' of 30 Squadron over Hinaidi, 1928. Note under-wing light bomb rack attached to 'spares container' basket.

Tight trio of DH9As, 30 Squadron returning from a bombing raid on Sulamania, 1924. Nearest, J7124, piloted by Flt Lt S M Kinkead, had a hung-up 112lb bomb still aboard but he landed safely; while the leading (centre) Ninak was piloted by Sqn Ldr J M Robb, later ACM Sir James, GCB,KBE,DSO,DFC. At this time 30 Sqn had aircraft wing-tips painted in Flight colours i.e. A Flt-Red, B Flt-Blue, and C Flt-Black.

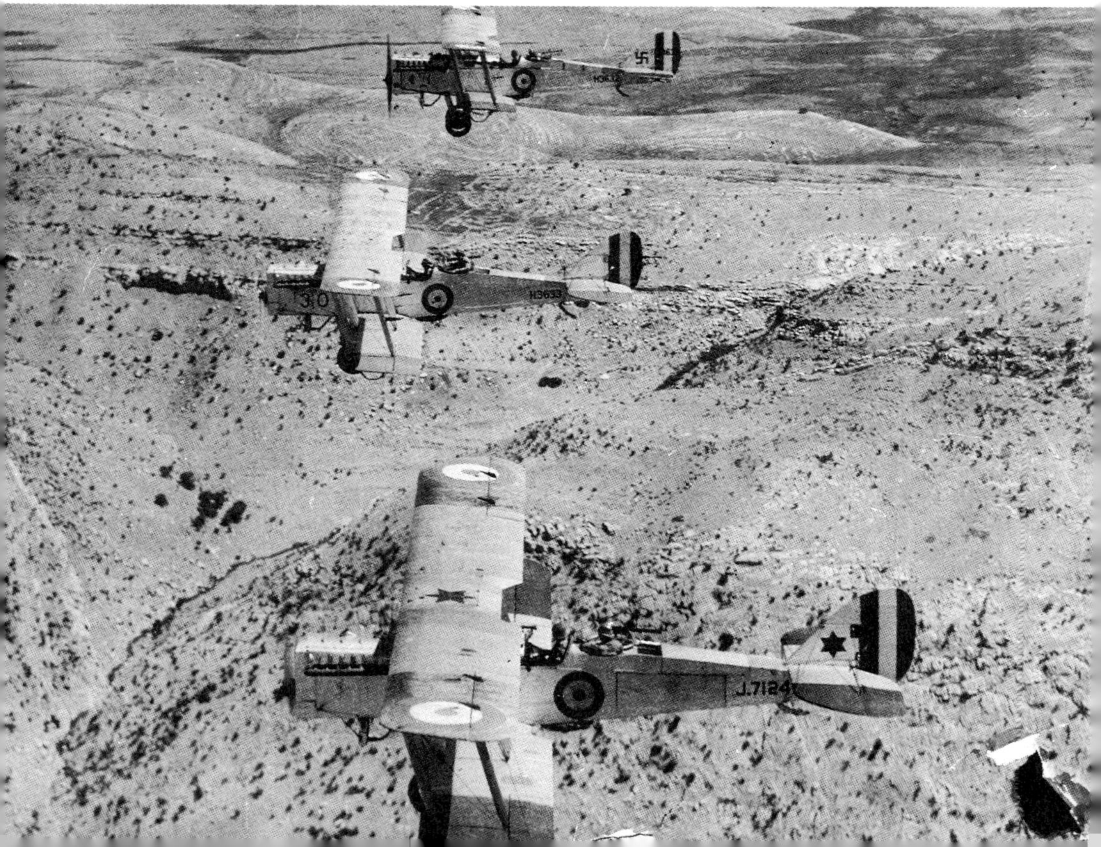

operations immediately by relieving 6 Squadron of its task of co-operating with the Samawah garrison from Baghdad, then in the following month moved base to Shaibah which was to become 84's spiritual home for the next two decades. In September 1920 too No 55 Squadron, reformed at Suez on 1 February 1920 with DH9s and DH9As initially and serving in Turkey by July, moved to Baghdad to replace 84 Squadron. Thus, by December 1920, operations flown by all four squadrons since June amounted to:

Total hours flown	4008
Total weight of bombs dropped	97½ tons
Total rounds SAA fired	183,861

RAF casualties for the same period, men and machines, amounted to:

Officers killed in action	6
Other Ranks killed in action	3
Officers wounded in action	5
Other Ranks wounded in action	2
Aeroplanes lost in hostile country	11
Aeroplanes put out of action by hostile rifle fire	57*

To add some scale to the intensity of the insurrection, overall army casualties from 2 July to 17 October 1920 alone amounted to 2,269 killed, wounded or missing; of which 312 were known killed and most of the 451 listed 'missing' were later confirmed as dead. Total Arab casualty figures were estimated as some 8,450 killed and wounded at least.

Conditions throughout Mesopotamia and its neighbours during 1921 were conducive to more peaceful activities, although not entirely free of operations for the RAF. In May 1921, for example, No 8 Squadron – reformed at Helwan,

* With rare exception, all were repaired and returned to operations.

Egypt on 18 October 1920 and based at Baghdad by January 1921 – used its DH9As intensively over a ten-day period, flying 300 hours and dropping 20 tons of bombs on rebelling Surehi tribesmen. Then from 14-16 June six 'Ninaks' of 8 Squadron joined with 55 Squadron in dropping ten tons of bombs on prevention of a Turkish incursion at Rowanduz – the beginning of anti-Turkish sorties until the end of September that year.

Yet despite the continuing operational status of the RAF in Mesopotamia *et al*, it was by no means confined to warlike activities; indeed, the greater part of RAF effort during the years 1920-30 in the Middle East zones concerned establishment of permanent communication routes, air-mapping of the bald desert wastelands, and pioneering long-range flights as precursors of future civil airline networks which would ultimately stretch from the United Kingdom via the Middle East to India and even Australia.

On 21 June 1921, DH9As from Nos 30 and 47 Squadrons* initiated the so-termed 'Cairo-Baghdad Mail Run' by air, while in the previous month these two squadrons had co-operated on opening up a permanent Cairo-Baghdad land route by supplying aerial escorts for vehicle convoys establishing a visible track across the desert. This latter primitive, but effective, route was officially open from 23 June 1921, and in 1922 the original track carved out of the sand was literally ploughed, by Fordson tractor, along its complete 840 miles length, with emergency landing grounds being marked every twenty miles as guide points for aircraft – the genesis of the projected 'All-Red' (i.e. British Empire) route to the Far East for civil air mail and transport.

Although such pioneering ventures as the Baghdad Mail Run may not be considered strictly operational actions, the hardships and perils of flying vintage aircraft over untracked

* No 47 Squadron had been reformed at Helwan, Egypt on 1 February 1920 – by renumbering No 206 Squadron – equipped with DH9As; although one Flight was almost immediately detached south to Khartoum and remained there until joined by the rest of the squadron in October 1927.

DH9As of 'C' Flight, 55 Squadron at Hinaidi, 1926.

Aerial view of No.4 FTS, Abu Sueir, Egypt in the 1930s.

Bristol F2B of 4 Squadron taking off from HMS *Argus*, 11 October 1922.

Bristol F2Bs of No.4 Squadron at Kilid Bahr, Gallipoli in January 1923, well muffled against appalling wintry conditions. Centre foreground is J6704, while at rt, 'A1' was J6715.

Bristol F2Bs, including H1623 & H1501, of 208 Squadron at San Stefano, Turkey.

tribal territories were no less than any straightforward bombing sortie.*

The combination of unrest and revolutionary uprisings during 1920-21 in Mesopotamia, Persia, Kurdistan and other adjacent areas, with a looming threat of further menace from the new Turkish nation, now left an anxious British government facing a serious dilemma in the context of its commitments to all of its mandated territories. There appeared to be two main alternatives. On the one hand a large reinforcement of all overseas army garrisons, as suggested by the War Office, which would inevitably be enormously expensive and, in view of Britain's post-1918 severe retrenchment on all military expenditure, would become a political albatross. Or, as was actually proposed by certain senior government figures, complete British withdrawal from such contentious areas as Palestine – a move which could surely lead to massive bloodshed when unrestrained Arab hatred of the new Jewish 'state' then followed.

As a result the Cairo Conference of March 1921 was convened to seek a solution, from which emerged Churchill's approval and support for Hugh Trenchard's proposal for air control as a substitution for army control of such territories; a scheme having the political blessing of costing only a fraction of the expense involved in maintaining any large army occupation forces. In pursuance of that decision it was agreed to increase RAF strength in Mesopotamia and bordering lands to a total of eight squadrons by the end of 1922. In addition *overall* command of British forces would then be invested in a senior RAF officer, and the first-ever such 'supremo' in Iraq† was Air Vice-Marshal Sir John Salmond who took up his appointment as GOC, Iraq with effect from 1 October 1922.

Before Salmond had taken up his new office, however, the relative peace of the Middle East was again disrupted, owing largely to the growing ambitions of a revitalised Turkey. After

* Readers are commended to the excellent firsthand account, *The Baghdad Air Mail* by Wing Commander R. Hill; Edward Arnold, 1929.

† As mentioned earlier, Mesopotamia was officially retitled Iraq in September 1921 as a first step towards the country's eventual independence.

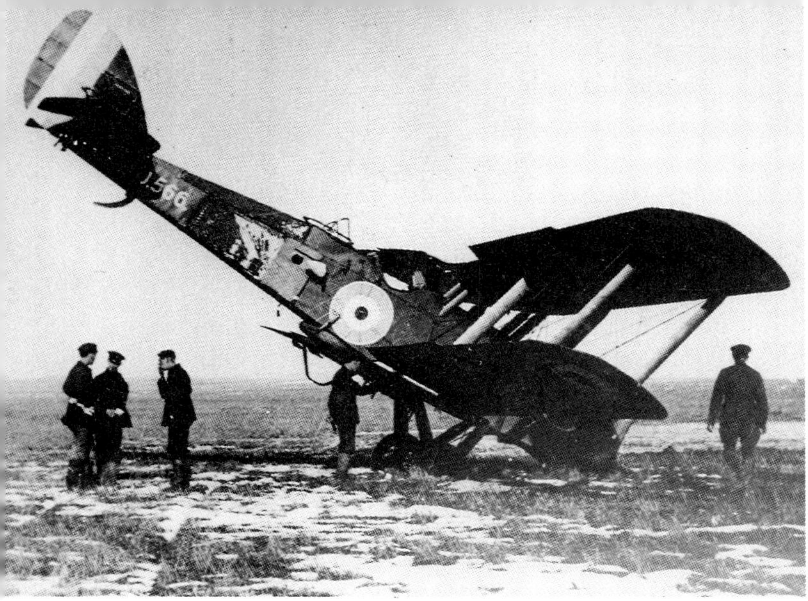

DH9A, J566 of 207 Squadron demonstrating a form of 'three-point' landing at San Stefano.

DH9As of 'A' Flt, 207 Squadron at Constantinople, including F2818 (nearest) and E8681.

Bristol F2B, J6789 of 208 Squadron, Turkey, 1923. Cartoon on fin appears to be a top-hatted puppy (?).

its humiliating defeat by the Allies in 1918, Turkey soon re-emerged as a new, energetic nation, hungry to reclaim the lands lost by the peace terms and determined to oust all foreign occupants of former Turkish possessions.

Chief architect of this nationalistic revival was one Mustapha Kemal Bey (later more generally named Kemal Ataturk). Son of a rug-maker, Kemal had been a professional soldier, partly responsible for the successful defence of the Dardanelles against wartime Allied attacks, and was now accepted by the Turks as virtual dictator. His despotic, hard-living nature had earned him the nickname of 'Grey Wolf', and, apart from his prime aim of regaining former Turkish territories, he was determined to westernise his country by such measures as freeing women from the harem, abolishing their veils, and introducing new educational systems for all.

While gathering a new Turkish army Kemal instigated unrest among various Kurdish tribes by promising help in ejecting their new British rulers, particularly in the vilayet of Mosul, then, once sufficiently armed in strength, began his campaigns against 'foreign invaders'. His first success of note came on 26 August 1922 when the Turkish army heavily defeated Greek 'occupation' forces at Anatolia.* Pursuing the survivors of this battle, the Turks reached Smyrna which they invested, burned to the ground, and then slaughtered thousands of its Christian inhabitants.

Continuing its triumphant advance Kemal's army now aimed towards its ultimate goal, occupation of Constantinople (now Istanbul), and by mid-September 1922 had reached and encamped near Chanak, just inside the Allied neutral zone around Gallipoli. This neutral zone was manned by token forces from Britain, France and Italy – very much in the nature of similar 'neutral' United Nations' forces of more recent years – which could do little to thwart any determined invasion and occupation of the zone.

The sudden appearance of Kemal's force on the borders of the strips of land on either side of the Gallipoli Straits which

* Under the Sèvres Treaty, 1920 Greece had been given the mandates of Smyrna and district and eastern Thrace.

had been designated neutral in the Peace Treaty of Sèvres created immediate tensions in the various European government circles. Since Kemal had already loudly repudiated that Treaty as a blemish on Turkish national honour, and now appeared determined to re-occupy the old Turkish capital of Constantinople, it presented the distinct probability that Kemal was fully prepared to sweep aside all foreign forces within said neutral zone in order to attain his objective.

Britain, among others, stood on the brink of war with Turkey. Hasty conferences during the week-end of 15-17 September in London resulted in Lloyd George's government resolving to hold Chanak and to stand fast against any Turkish incursion. With the French and Italian forces about to be withdrawn – the French government had secretly, even deviously backed and supplied the Turks, while Italy wanted no part of any full-scale conflict with Kemal – the British garrison there, commanded by General Sir Charles H. Harrington, stood alone in Kemal's path to Constantinople, and clearly required heavy reinforcement – and quickly.

While Kemal's force remained virtually surrounding the barbed wire perimeter of the British position at Chanak, various warnings not to violate the neutrality of the Gallipoli zones were despatched to the Turkish commander – who calmly replied that he recognised no such zones. Meanwhile air, land and naval forces were being assembled and despatched to the area from Britain and other Middle East sources. One RAF troopship, *Braemar Castle*, laden with a full complement of RAF personnel and supplies intended for Iraq, was diverted to Chanak where the RAF element was disembarked and placed at General Harrington's disposal; while the seaplane carrier HMS *Pegasus*, already in the nearby Mediterranean, conveyed further troops to Chanak on 9 September, then tasked its own complement of five Fairey IIID floatplanes with daily reconnaissance patrols over the Turkish positions, gathering information. In Britain mobilisation orders were issued to certain RAF squadrons based in the UK and in Egypt to prepare for an immediate move to the Dardanelles. Of these, No 203 Squadron aboard the aircraft carrier HMS *Argus*,

Nieuport Nightjar at San Stefano.

Servicing a Neuport Nightjar of 203 Squadron at Kilia, 1923.

comprised of five Nieuport Nightjar single-seat carrier fighters and eleven Fairey IIIDs, left Portsmouth on 18 September, docked briefly at Malta to embark six more IIIDs, then arrived at Kilia, directly across the Straits from Chanak, where all aircraft were off-loaded and commenced regular recce patrols over the Turkish dispositions. Three other UK-based squadrons – Nos 4 (Bristol F2Bs, Farnborough), 25 (Sopwith Snipes, Hawkinge), and 207 (DH9As, Bircham Newton) – also prepared quickly, while from Egypt were to come Nos 208 (Bristol F2Bs), 56 (one Flight of Snipes), apart from 16 Fairey IIIDs of 267 Squadron based at Kalafrana, Malta.

While awaiting the arrival of all these RAF units in late September-early October, General Harrington at his HQ in Constantinople could only pursue his hope for a peaceful solution to the crisis. Personally convinced that Kemal would not risk an all-out confrontation with Britain, he offered to enter arbitration with Kemal's representatives at Mudania in early October, turning a Nelsonian blind eye to a definite instruction received on 29 September from the UK government – instigated primarily by the ever-pugnacious Winston Churchill – to issue an ultimatum to Kemel to withdraw all his forces, failing which all British forces would 'open fire' (*sic*). In this interim period the RAF units began reaching the disputed zone, arriving initially in the following order:

Date	*Squadron*	*Aircraft*	*Commanding Officer*
26/9/22	267	Fairey IIID	Flt Lt J.A. Sadler
27/9/22	203	Nightjar	Flt Lt L.H. Slatter, OBE,DSC,DFC
28/9/22	208	Bristol F2B	Sqn Ldr A. ap Ellis
28/9/22	56	Snipe	Flt Lt G.H.D. Gossip
8/10/22	4	Bristol F2B	Sqn Ldr C.H.B. Blount, MC
11/10/22	207	DH9A	Sqn Ldr A.W. Tedder
12/10/22	25	Snipe	Sqn Ldr Sir N. Leslie, Bt

All units now became generically titled Constantinople Wing, headquarters for which was established in Constantinople city, commanded by Group Captain R.M. Fellowes, DSO, while an

Aircraft Park was set up at San Stefano airfield, some seven miles south-west of the city.

Necessarily, all squadron aircraft and personnel had been conveyed to the Dardanelles by sea under varying conditions. No 4 Squadron, for example, had embarked on the seaplane-carrier HMS *Ark Royal* and arrived in Kilia Bay on 8 October, then had to have its Bristol F2Bs transferred to the aircraft carrier *Argus*. On 11 October all twelve squadron crews flew their F2Bs off *Argus*'s deck and landed on Kilia airfield without mishap, although it had been the first time any of the pilots had ever taken off from a carrier at sea.

Its companion F2B unit, 208 Squadron, based at Ismailia, Egypt, had been transported aboard SS *Podesta* which docked at Constantinople from where 208 and the 56 Squadron Snipe Flight were moved to San Stefano. Three days later these latter units commenced recce patrols in conjunction with the other squadrons. No 207 Squadron's advance party and then the main unit left England aboard the ships SS *Eboe* and *Khartoum* and reached Constantinople on 11 October, but were left aboard temporarily before finally being off-loaded to move to San Stefano.

All units arrived at a time when local weather conditions were deteriorating rapidly with the onset of winter. At San Stefano the airfield was in a shocking state from floods and frost, and 207 Squadron's DH9As, picketed in the open due to the unsuitability of the obsolete canvas hangars issued, suffered badly. With no fresh dope available fabric coverings soon split, while airscrew blades suffered splitting and cracking, necessitating exchange after as little as five hours' flying.

No 4 Squadron at Kilia fared no better, having to cope with violent dust-storms and extreme heat conditions during its first two weeks, then enduring the advent of the rain season which quickly transformed the airfield into a quagmire from which the F2Bs were unable to take off, or even taxi without sinking axle-deep in the glutinous mud. Later in October the squadron was to suffer further when 60-70 mph gales wrecked hangars and accommodation – all tented – and these were followed by floods. All attempts by 4 Squadron to improve the airfield state failed and by January 1923 the unit gave up the struggle and moved en bloc to Kilid el Bahr where all aircraft continued to be maintained in the open, there being no hangars available.

One of the very few 'light' moments of that demoralising period came when one of the portable canvas hangars issued to 4 Squadron in England from wartime storage was unpacked prior to initial assembly. Inside were found some six-year-old letters and parcels, dated 1916, with one letter signed by 'Two Old Ladies of Ascot'.

While the squadrons began settling in to their new venues, Sir Charles Harrington had continued attempting to reach peaceful agreement with Kemal's envoy, Ismet Pasha at Mudania, with little success. Then, on 10 October Harrington received a signal from his local commander Major-General Marden stating that the Turks had commenced enveloping the British garrison force at Chanak and requesting permission to open fire in 'self-defence'.

Harrington replied immediately with that permission but deferred zero hour for same until later that day. He then made one final attempt to persuade Ismet Pasha, emphasising that he would not surrender Chanak at any cost. To Harrington's utter surprise the Turk suddenly agreed and Harrington was able to cancel his order to Marden – a mere 75 minutes before his stated zero hour. Next day a peace pact was duly signed at Mudania by both sides, although this had yet to be properly ratified by an eventual true peace treaty. For the moment at least all-out war with Turkey had been avoided.

Notwithstanding the Mudania peace pact, Harrington remained suspicious of Turkish intentions and ordered continuance of the regular aerial reconnaissance flights by all RAF units, and retained the Constantinople Wing in situ until the eventual formal peace treaty was signed and authorised in Lausanne on 23 August 1923. For the RAF squadrons the better weather of spring and summer 1923 offered the opportunity to undertake various training programmes of bombing, gunnery, and general army/navy co-operation exercises and trials, but once the Lausanne Treaty had been signed all units began returning to their various UK and

Egyptian bases of the previous year.

The draft of airmen originally intended for Iraq, which had been diverted to Constantinople, provided a final flourish by parading fúlly booted and spurred through the city streets prior to continuing its interrupted transit to Iraq. This parade also provoked an ironic note of humour when an English female spectator on a nearby balcony was distinctly heard to say, 'Smart, aren't they? But how in God's name did the Portuguese in those adorable blue uniforms get mixed up in this mess?!'

For the RAF, the Chanak Crisis had been its first real postwar test of its capability to mobilise and transfer a fighting formation to any foreign troublespot in short time (and, indeed, at virtual over-night notice), and the experience taught valuable lessons for future eventualities of any similar nature. Every squadron commander involved could testify to the abysmal organisation of the UK stores depots in their equipping the mobilised units initially, while equally awful disorganisation at the receiving end in the Dardanelles had merely added another catalogue of difficulties. Yet the air and ground crews had buckled down and produced their aircraft despite appalling living and working conditions – a mark of the growing spirit of the 'Junior Service' as it very slowly matured.

CHAPTER FIVE

Mahmud

When, on 1 October 1922, Air Vice-Marshal Sir John Salmond officially became GOC of all British forces in Iraq, among them were eight RAF squadrons based in Iraq:

Squadron	*Base*	*Aircraft*
1	Hinaidi	Sopwith Snipe
6	Mosul	Bristol F2B
8	Hinaidi	DH9A
30	Baghdad West	DH9A
45	Hinaidi	Vickers Vernon
55	Mosul	DH9A
70	Baghdad West	Vickers Vernon
84	Shaibah	DH9A

Of these, No 1 Squadron had been transferred from India in April 1922, while Nos 45 and 70 had moved to Iraq from Egypt in early 1922; all three movements being in accord with the decision of the Cairo Conference of 1921 to raise the RAF Iraq establishment to eight squadrons.

On taking up his appointment Salmond inherited a number of immediate problems in the contexts of tribal and other insurrections. The chief menace to peace within his domain was Turkey, as Mustapha Kemal began his self-imposed crusade to regain control of all former Turkish possessions, including Kurdistan and other areas of northern Iraq. Apart from the main Turkish thrusts to oust Greek forces from their mandated territories, coupled with the aim of re-occupying the old Turkish capital city, Constantinople (see previous chapter),

Turks were infiltrating various northern Iraqi tribal lands, providing dissidents with arms and ammunition and promising aid for any anti-British revolts.

Entwined with this Turkish covert movement was Sheikh Mahmud who, in 1922, had been released from prison by the British* and appointed as governor of the Sulaimania district when British administration was withdrawn from that area. Mahmud almost immediately began mounting an anti-British revolt, with Turkish backing.

By February 1923 British intelligence confirmed that Mahmud was planning an all-out attack on Kirkuk as the spark to a general uprising of the Kurdish tribes; while a Turkish despatch was also intercepted which told of intended attacks on Kirkuk, Erbil and Koi Sanjak, allied to Mahmud's rebellion. In view of this intelligence Mahmud was summoned to Baghdad but ignored the summons, and air action – dropping delayed action bombs around Sulaimania at first, then bombing the sheikh's own quarters in the town – was promptly undertaken. Mahmud merely fled to cave refuges in the surrounding hills and continued his campaign plans. Accordingly, a combined air and ground operation was mounted – two columns of Imperial and Iraq Levies' troops respectively, supported by RAF detachments based at Mosul, Kirkuk, and Erbil.

By April the ground forces had occupied Rowanduz, and by the end of that month had forced all Turkish forces back over the border of Persia. These columns next advanced on Sulaimania which was finally occupied on 17 May, forcing Mahmud and his personal entourage to take temporary refuge in Persia. Throughout these operations the RAF support units carried out daily reconnaissance and bombing sorties, but also air-supplied rations, stores and clothing *et al*, while the Vernon aircraft (mainly) evacuated sick and wounded troops from otherwise inaccessible locations.

While these operations were being conducted in the northern areas trouble erupted in part of southern Iraq, where a number of dissident sheikhs thought the time ripe to revolt

* In June 1922, to live in Kuwait with an allowance of 2,000 rupees per month!

RAF Hinaidi, principal base for the RAF Iraq in the 1920s. The hangars at far end housed Nos 45 and 70 Squadrons.

Vickers Vernon ambulance (J7143) of 45 Squadron over Hinaidi, piloted by Flg Off B Embry (later ACM Sir Basil). Note uncowled engines.

while the British were – apparently – well occupied with Mahmud and the Turks. They were to be swiftly disillusioned. Vernon aircraft took on the role of bombers and in just nine individual sorties 'delivered' some 6½ tons of high explosive bombs, 4,000 incendiary bombs, and fired almost 2,000 rounds of Lewis gun ammunition in routing the tribesmen. In command of No 45 Squadron at this period was Squadron Leader Arthur Harris, whose Flight commanders were Flight Lieutenants R.H.M.S. Saundby and R.A. Cochrane, and at Harris's suggestion several of the squadron's Vernon troop-carrier aircraft had had holes sawn in the noses for bomb-aiming, and bomb carriers fitted under the lower wings; a totally unofficial modification but characteristic of Harris's pragmatic view of his duties. Harris eventually commanded RAF Bomber Command from 1942-45, with Saundby as his deputy, and Ralph Cochrane as AOC-in-C, No 5 Group. Harris's personal memories of his two years in Iraq (1922-24) were not fragrant:

> Never shall I forget the appalling climate, the filthy food, and the ghastly lack of every sort of amenity that our unfortunate men were compelled to put up with in 'peace time', for two years at a stretch,* away from their families. A main result of the economy campaign in the home newspapers was to deprive us of the most meagre and essential comforts.†

A constant corollary to any air operations was the distinct possibility of having to force-land in tribal territory due to engine failure, other mechanical troubles, or the not uncommon result of very accurate rifle-fire from tribesmen being attacked. On numerous occasions of that ilk, it became accepted practice for companion aircraft to land alongside the crippled machine, then 'lift' the stranded crew back to base, despite often being surrounded by tribesmen hotly intent

* A reference to the contemporary custom of airmen posted overseas for their five years (minimum) tour of duty often 'splitting' the tour by two years in Iraq and three in India.

† *Bomber Offensive* by Sir A. Harris; Collins, 1947.

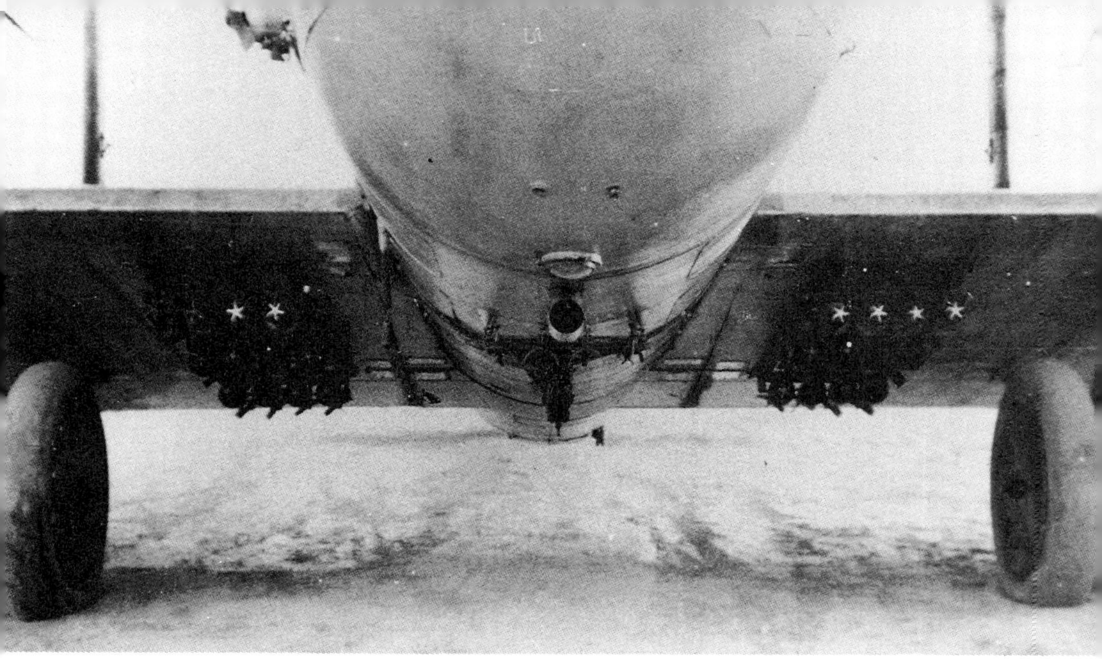

Vernon of 45 Squadron loaded with 25lb Cooper bombs.

Vernon '*Pelican*' of 45 Squadron, with King Feisal of Iraq in cockpit. Other squadron Vernons were named after Kipling's mine-sweepers e.g. *Unity*, *Claribel*, *Assyrian*, *Stormcock*, *Golden Gain*.

on killing or capturing the crews and their equipment.

Though all crews carried their 'goolie' chits, these were by no means guarantees of safety, and captivity by rebelling Arabs or Kurds too often meant torture, mutilation, and death. One DH9A crew from No 8 Squadron in 1923 faced such a prospect for over an hour before finally being retrieved. Piloted by Pilot Officer Neville Vintcent, with Flight Lieutenant J.I.T. Jones, DSO,MC,DFC,MM as his passenger, the Ninak set out in company to attack rebel Arabs at Samawah, duly bombed and strafed the target, but then had engine failure resulting in a forced landing while strafing some tribesmen about two miles from Samawah. Firing Very lights to attract the attention of the other bombers, Vintcent and Jones were dismayed to get no response, but then a Snipe piloted by F.L. Luxmore, DFC, followed shortly after by a DH9A, piloted by Flight Lieutenant Jack Cottle, DFC (later Group Captain, MBE,DFC), both landed nearby and offered to lift the stranded crew off. Since there was only room for one extra man on Cottle's DH9A, Vintcent and Jones elected to stay with their machine while Cottle went to fetch some petrol. As soon as Cottle and Luxmore had left, tribesmen began sniping at the grounded Ninak from close range. Jones replied with brief bursts from the rear-seat Lewis gun, and when any threatened attack came from any direction 'blind' to Jones' arc of fire, Vintcent coolly got his shoulders under the Ninak's rear fuselage and heaved it round to give Jones a clear field of fire. This procedure was carried out for almost an hour before four aircraft finally appeared, landed, and retrieved the crew. Vintcent, who had only recently graduated from the RAF College at Cranwell, was awarded a DFC – the first ex-Cranwell cadet to receive this decoration.

Sheikh Mahmud's latest sojourn in Persia was brief and in the summer of 1923 he returned to Sulaimania and proclaimed himself 'King of Kurdistan', obviously still intent on ridding 'his' country of all foreign control (albeit still accepting secret aid from the Turks). The British authorities, rather than mount yet another punitive expedition, offered Mahmud a concession that providing he restricted his activities within a defined area he would not be molested. Mahmud's reaction was predictable – he continued to advocate rebellion. Air action followed but failed to quell Mahmud's ambition.

Then in early May 1924 a minor clash between some Assyrian Levies and Moslems in Kirkuk – many of the latter being killed – provided Mahmud with a slender excuse for declaring a *Jihad* (Holy War) against the British and the Assyrians, and immediately set about gathering a large *lashkar* (armed force) for this purpose.

To counter this latest threat the British authorities sent a force of 42 aircraft drawn from Nos 6 and 8 Squadrons – detached to Kingerban – and Nos 30, 45, 55 and 70 Squadrons – based at Kirkuk – which first issued proclamations that Sulaimania would be bombed if Mahmud refused to submit. Mahmud, as usual, ignored the warning, and the squadrons proceeded to drop 28 tons of bombs on Sulaimania during an unceasing 48-hours' period, destroying much of the town, although its inhabitants had already abandoned their dwellings and sought refuge in the hills. In July Sulaimania was re-occupied by British forces and, with the onset of winter conditions, relative peace was established in the area.

Almost immediately the scene of operations shifted back to the Mosul district where Turkey renewed her attempts to occupy the Mosul vilayet. On 13 September 1924 intelligence was received indicating that Turkish forces intended crossing the frontier to seize Zakho that same night. Nine Bristol F2Bs of 6 Squadron took off at dawn on 14 September and within fifteen minutes came upon some 300 Turkish cavalry crossing the river Borana. Sweeping into action, the F2B crews swiftly dispersed these troops at a cost of one pilot wounded by ground fire. Refuelled and re-armed, the F2Bs returned and effectively strafed and bombed a second Turkish cavalry force of some 150 men at Birkar.

Such prompt action served to stultify Turkish movement into Iraq for the moment, although further attempts were to be made before an eventual League of Nations initiative resulted in a tripartite peace agreement in 1926 at Angora. For one F2B pilot the first attack on 14 September had a grim overtone. As he swept in to strafe at zero height his starboard lower wing-tip

Sopwith Snipes of No.1 Squadron over River Tigris, Baghdad, 1924.

DH9As of No.8 Squadron, 1924 at Hinaidi, with (nearest) Nieuport Nighthawk attached for 'tropical trials'.

Neat line-abreast trio of Westland Wapitis of No.84 Squadron over typical bald Iraqi desert landscape.

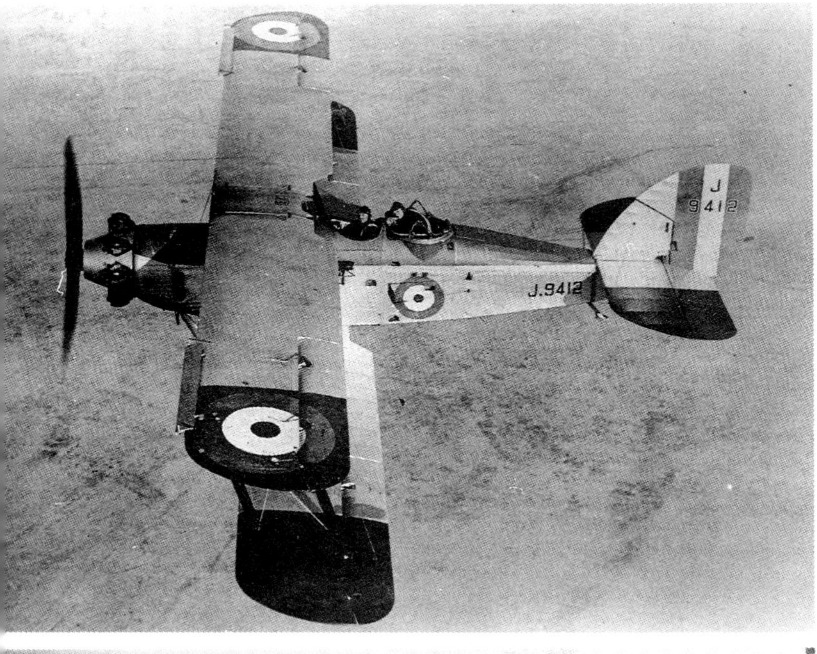

Close-up view of Wapiti J9412, 30 Squadron.

Wapiti K1124, 'B3' of 55 Squadron, bearing unit badg on its fin.

caught one unfortunate mounted Turk literally head-on and the F2B's rear gunner reported after that the Turk had been decapitated. The pilot, totally absorbed in sighting for his forward machine gun, felt no impact and only believed his observer when he was shown the bloody and bent wing-tip later ...

The 1924-25 winter 'peace' soon evaporated in the spring of 1925 with Mahmud once more stirring up revolt, his forces attacking a Levy column of cavalry moving towards occupation of Halebja. The ensuing battle began reaching serious level but at a critical point of the clashes four Sopwith Snipes of 1 Squadron, detached to Sulaimania, caught a large body of Mahmud's men in open country and proceeded to annihilate this force by strafing – an action which enabled the Levis to occupy Halebja next day. Other air action against this latest Mahmud uprising was undertaken by 6 Squadron's F2Bs and the DH9As of Nos 30 and 55 Squadrons. In early June 1925 British ground troops occupied Choartah, the principal village of the Sharbazher district where Mahmud's influence over the tribes was normally undisputed. On 20 June Mahmud instigated an all-out attack on Choartah only to be heavily defeated by prompt air action in which 6 Squadron figured prominently, leaving Mahmud to retreat to his Persian bolt-hole once more, and apart from one more nullified attempt in September he was to remain inactive for the remainder of that year.

Mahmud was to remain inactive during the winter 1925-26 but in the spring of 1926 resumed his agitations by despatching small parties of rebels to infiltrate various tribal villages where they took over control. Bad weather conditions in March-April prevented any attempts by British forces to oust these rebels, and by June full-scale land and air action became possible, leading to severe fighting. On 17-19 June intensive air action against rebel villages was taken by 30 Squadron's DH9As, based at Kirkuk, and Snipes of 1 Squadron, detached at Sulaimania*,

* One of 1 Squadron's pilots in these operations was Flying Officer D. Boyle, later to become MRAF Sir Dermot Boyle, GCB,KCVO,KBE,AFC, Chief of the Air Staff, 1956-59.

which assisted greatly in finally dispersing Mahmud's men. Mahmud, once again, 'retired' to his customary refuge in Persia, but took with him a DH9A crew, forced down with engine trouble on 14 June. These prisoners were eventually returned in October, and early in 1927 Mahmud was persuaded to visit Baghdad where he eventually agreed to terms and withdrew from activities for the next three years. With Mahmud muted, a decision was taken to strengthen government forces in another rebel leader's area of influence in 1927, when long simmering unrest among the tribes of central Kurdistan was being brought to boiling point by Sheikh Ahmad of Barzan. Ahmad had been hostile to all foreign administration control since 1919 and had escalated his covert opposition in 1925 when his area came under nominal control of the Iraqi government. Thus in 1927 a force of Iraq Levies, supported by 6 Squadron and a Flight of 55 Squadron's DH9As occupied Barzan, causing Ahmad and his 'court' to flee to the hills.

No 1 Squadron continued to maintain a detachment of Snipes at Sulaimania until September 1927, but was then informed that the squadron would be locally disbanded on 1 November, then reformed as a fighter unit in England shortly after. Its Snipes did not have the range for the normal forms of operations against hill tribesmen, while the total lack of aerial opposition obviated any form of 'fighter escort' role. Squadron personnel were, in a majority, simply reposted to other Iraq units to complete their tours, while all the Snipes were scrapped in situ. Among the units to receive ex-1 Squadron pilots was No 6, based at Mosul, commanded by Squadron Leader C.N. Lowe, MC,DFC (later, Group Captain).

The only Bristol F2B unit based in Iraq – normal aircraft establishment being twelve F2Bs – 6 Squadron received a new commander, Squadron Leader C.H. Keith, in January 1928, whose first impressions of his new command were:

> My command is within a barbed wire perimeter camp – with the exception of the water-pumping and purifying station and the ice plant, which are as near as possible on the right

RACING RED NINE. DH9A, J6959, painted overall blood red with white strutting; the personal hack of AVM Sir Edward Ellington, 1928.

VIP WAPITI. The all-white personal aircraft (J9592) of AVM E. Ludlow-Hewitt at Hinaidi, 1932 on relinquishment of his command as AOC, Iraq.

> bank of the Tigris. The place is intended to be defensible, and I am supposed to have, within the barbed wire, all the necessities of life to hold out, if surrounded. There is a complete army co-operation squadron (No 6) of Bristol Fighter aircraft, No 8 of our Armoured Car Company, a Supply Depot, a wireless station and a sort of small hospital. The station comprises a great number and variety of complicated and dilapidated buildings, which seem to have been put up without any definite plan ... a block-house, now being used as a turkey pen, is destined to become a camera-obscura hut! The old Mess dining room was formerly the swimming bath in the harem quarters of an old Turkish house ... The job, as a job, is a good one, and I am almighty glad to get it. It is probably the most independent squadron leader's command in the entire RAF, and my aircraft are responsible for all necessary action within an area as big as all England.*

Keith's comment on the 'independence' of his job as a squadron commander was applicable in some degree to every RAF unit CO. With the constant need for detachments to bare desert landing grounds for active operations, even very junior officers were expected to assume personal responsibility for decisions and actions usually requiring far senior rank and authority; experience which tested severely the 'leadership' qualities of young officers in that era, but which would be an invaluable foundation for later high command in many cases.

While Sheikh Mahmud was to create no further problems until 1930, the British and Iraqi authorities meanwhile had plentiful problems in the context of maintaining peace and the law with various tribes. These troubles arose from different tribes vying for supremacy over other tribes and their territories, particularly encouraged by such prominent Arabs as Ibn Saud of the Nejd territory which bordered Kuwait and Iraq in the southern areas. These inter-tribal attacks and unrest

* *Flying Years* by C.H. Keith: J. Hamilton, 1937.

Wapiti J9835 of 84 Squadron, Shaibah, March 1931, bearing the unit's scorpion badge.

SHOOTING THE NULLAH. Wapiti K1391 of 55 Squadron near Mosul, 1930, easing through the mountain valleys.

were not new, having occurred spas- modically since 1920-21, but by late 1927 tensions had increased to a point where, on the night of 5/6 November, a party of Mutair tribesmen attacked a newly-created Iraqi camel-police post at Busaiya, killing all occupants except one policeman who escaped to report the massacre, and a wounded man discovered next day. The perpetrators fled beyond the frontier, safe in Nedj territory.

Within the following month it became quite obvious that Ibn Saud was either unwilling or unable to prevent his tribes from their continuing raids across the Iraq border – in view of his personal hostility towards King Feisal of Iraq it was most likely a case of Ibn Saud turning a blind eye to such incursions – and he was given a specific warning to restrain his men within six weeks or Iraq forces would 'blockade' (*sic*) the Nejd and Hasa coasts, besides taking air action beyond the frontier against Nejd tribes; such latter action being authorised on 3 January 1928.

In anticipation of failure by Ibn Saud to comply with this warning an Advanced Headquarters, titled 'Akforce', was formed on 8 January at Ur Junction, including elements of Nos 55, 70 and 84 Squadrons and sections of the Armoured Car Wing. Advance operational bases were established at Busaiya (nine DH9As of 55 Squadron plus three RAF armoured cars) and Sulman (nine DH9As of 84 Squadron plus three armoured cars) with twelve Victorias and Vernons based at Akforce HQ for back-up of air supplies *et al.*

On 11 January warning notices were dropped in Nejd territory telling the tribesmen to ' ... retreat to a distance of four days' march from the Iraq frontier' (*sic*), failing which any persons found still in this proscribed area would be liable to air attack without further warning. At first the Nejdi tribes concerned seemed ready to comply but soon had to be 'encouraged' by the dropping of 'warning' bombs around their encampments, though this bombing caused no casualties. On 27 January, however, a party of Akhwan some 300-strong raided Kuwait tribes some 70 miles south-west of Basra, and next day RAF aircraft joined in the pursuit of the raiders, catching one large party fifteen miles north-east of Hafar and attacking same in the face of combined and fierce rifle-fire, but losing one aircraft which force-landed though its crew were rescued by their companions.

Next morning two formations of six aircraft each set out again and located a party of raiders eight miles west of Hafar. During the ensuing attack one DH9A, shot through its radiator, force-landed merely 400 yards from the raiders, but its pilot (who was alone) was retrieved by Flight Lieutenant J.F.T. Barrett, DFC* in another Ninak despite being under a hail of fire. Barrett was subsequently awarded a DSO for this and other actions during these operations.

By February permission had been obtained for the use of Kuwait bases for any further operations when needed, which were soon used to repel intended raids by Akhwan tribes. By now the gross total of rebel tribesmen was estimated at some 50,000, although individual raiding parties were usually no more than 4000 at most.

At the end of February Akforce was reinforced by a Flight from 30 Squadron plus another section of armoured cars from Hinaidi, while arrangements were made for one Flight of Victorias of 216 Squadron, based at Heliopolis, Egypt to be held in readiness for a move to Iraq if needed. March and April saw all aircraft engaged in ceaseless recce patrols but few actual air attacks, there being growing evidence that many tribes were retreating inland and ceasing raiding.

In May, however, 84 Squadron's DH9A crews carried out a unique raid when they bombed the remote island of Gubbah, on the Hammar Lake, which had become one rebel sheikh's stronghold. The Ninaks' bombs were deliberately dropped on the *bund* (banking) which surrounded the island, thereby flooding the rebels out of their refuge – an operation pre-dating the legendary 617 Squadron 'Dam-busting' operation of 1943 by fifteen years! In the same month Ibn Saud finally gave a written undertaking to restrain his tribes from raiding and by 3 June Akforce had ceased to exist as all units withdrew to their normal bases and duties.

* Later Wing Commander, DSO,DFC.

Throughout these anti-Nejd operations aircraft had only flown actual bombing sorties on 26 separate occasions, using mainly 25lb Cooper bombs and a few 520lb, but the intensity of overall flying involved is illustrated by each unit's total flying hours:

Month	*30 Sqn*	*55 Sqn*	*70 Sqn*	*84 Sqn*
January	–	441	406	542
February	–	473	474	544
March	–	588	445	578
April	337	636	499	475
May	–	270	242	326
	337	2408	2066	2465

Several aircraft had been brought down during operations, mainly due to mechanical problems, but only one DH9A pilot (Flying Officer Jackson) had been shot down and killed, on 20 February. His body was later recovered and conveyed to Basra for proper funeral arrangements.

These anti-Nejd operations were reasonably typical of the RAF's duties in Iraq and the Middle East generally during the 1920s. Flying constantly over trackless desert wastes of which few (if any) accurate maps existed then, living and working in primitive conditions for weeks, even months, at a stretch, ever-conscious of probable fates should engine failure occur while over tribal territories, yet never failing to fulfil given duties if humanly possible.

No small contribution to the air crews' daily hazards was the vintage designs of aircraft they continued to fly in the late 1920s and early 1930s. The stalwart aircraft of that period were the De Havilland 9A and its doughty companion, the Bristol F2B. Each had been designed during World War One, purely for that war's exigencies, and by the late 1920s were still virtually unchanged in performance and ability. In Iraq the DH9A was the prime workhorse and came to be steadily encumbered with a variety of excrescences – spare wheels, tropical radiators, bed-rolls and other personal crew equipment when on detachments, animal skin water-bags, external bomb carriers, *et al* – all of which merely diminished overall performance. Further extra loads for routine duties often included the bulky hand cameras of the period, necessary for the unending tasks of reconnaissance and aerial mapping, while a maximum bomb load under wings and fuselage could amount to 740lb, apart from the nominal two machine guns, spare ammunition, and – in later years – some aircraft fitted experimentally with wireless sets and their associated paraphernalia of batteries, W/T masts and the like.

For many pilots, fresh from the 'modern' air force equipment in the United Kingdom, flying such out-dated aircraft came as something of a shock, while the myriad difficulties of merely navigating across the Iraqi deserts where even normally easily recognisable features could be rendered 'invisible' within hours by changing light conditions and ever-shifting dust and sandstorms, meant that most new pilots were considered to need about a year's practical experience before being regarded as capable of undertaking fully independent sorties without the risk of getting lost. Such experience could only be gained the hard way – as one ex-6 Squadron Bristol F2B pilot summed up his own period in Iraq: 'I learned more about handling aircraft, and my own limitations, in 18 months on "Shiny Six" than in ten years of Service flying in England.'

By 1928 the obsolescence of the DH9A was finally recognised by higher authority and, slowly, replacement designs began reaching some units, although this process was to extend until 1931 before the last Ninak had left the Iraq scene. No 84 Squadron at Shaibah began its conversion to Westland Wapiti aircraft in that year and had two Flights equipped with Wapitis by mid-September, the squadron's last DH9A leaving in January 1929. The Wapitis issued to 84 Squadron were the first to enter RAF squadron service, and though designed as a replacement for DH9As and Bristol F2Bs, utilised certain DH9A components and spares – due partly to the prevailing economy policy of the UK government in the 1920s – but were

RAF station Moascar, Ismailia in Egypt, 1924.

THE OLD ORDER CHANGETH. Bristol F2B (FR4744) alongside its successor Fairey Gordon K1730, No.6 Squadron, at Ismailia, April 1932. The airship mooring tower at left had been built in anticipation of a projected 'all-red route' to India from the UK by rigid airships.

sturdier in construction and possessed a more powerful engine (Bristol Jupiter) which offered pilots an all-round improvement in flying performance. The other two Iraq-based DH9A units, Nos 30 and 55 Squadrons, began their conversions to Wapitis in 1930, although again the change came only slowly, with 55 Squadron still having a few DH9As on strength until early 1931.

Elsewhere in the Middle East No 14 Squadron in Palestine began re-equipment with Fairey IIIFs in December 1929, as did 45 Squadron at Helwan in September 1929, while 47 Squadron at Khartoum had already relinquished its DH9As for Fairey IIIFs in April 1928. Meanwhile, the remaining Bristol F2B squadrons in the Mediterranean zones – No 6 in Iraq and 208 in Egypt – had to soldier on with their ancient 'Brisfits'.

Nineteen twenty-nine, however, brought a complete change of scene for No 6 Squadron when, on 1 October, the unit despatched an advanced party of one officer and 25 airmen in five Victorias from 70 Squadron, bound ultimately for Moascar, Ismailia in Egypt – the start of a squadron move to its new base and finale to the unit's ten years of service in Iraq. Before leaving Mosul some of the unit's airmen carved out a large representation of the squadron's badge and motto – '*Oculi Exercitus*' ('The eyes of the Army') – on a huge rockface at Abu Saif, just south of their airfield, as a permanent memorial to the squadron's service there.

This squadron move became an experiment by the Air Staff in transference of an entire squadron's equipment and personnel from one command to another, able to arrive at a new base and resume firstline duties without delay. The land distance from Mosul to Moascar was slightly over 1,000 miles, all across deserts, and meant in 6 Squadron's case moving twelve aircraft and crews, twelve officers, 114 airmen and sufficient supplies in 24 Crossley Tenders by road and by air. The dozen F2Bs left Mosul by air on 2 November and arrived at Moascar two days later, while the road parties reached their destination by 11 November – all moves having no mishaps.

By mid-1930 all squadrons in Iraq had become involved in fairly routine duties, rather than actual operations, giving the air and ground crews reasonable time to adjust to their new aircraft, but in the autumn, after three years of silence, Sheikh Mahmud re-entered the area clearly bent on a final bid to establish his Kurdish kingdom. His hopes were partly based on disturbances among the tribes around Sulaimania on 6 September in connection with Iraqi parliamentary elections, and Mahmud began gathering an armed force on the Persian border. Warning letters from Iraqi high officials to Mahmud on 10 September were ignored, and on 17 September Mahmud crossed the border into Iraq, near Penjwin, north of Sulaimania, a district occupied by the Pizhder tribes. On 4 October the Iraq government ordered Mahmud to leave Iraqi territory 'forthwith' (*sic*), but he ignored this and the further ultimatum issued on 21 October.

Meanwhile an Iraqi Army force, comprised of a cavalry regiment, three infantry battalions and a mountain battery, was formed at Sulaimania, ready to move against Mahmud should he reject the ultimatum. Although the senior RAF commander was still GOC all forces in Iraq in 1930, he was conscious of the fact that the country would within two years achieve full independence, and accordingly it was decided that no Imperial troops would be used except in reserve, the whole anti-Mahmud campaign becoming wholly the responsibility of the Iraqi Army, albeit supported by the RAF as necessary.

To provide such RAF assistance a Flight of 30 Squadron was sent from Mosul to Sulaimania on 25 October to provide air reconnaissance for the army, while a composite 'squadron' consisting of one Flight of 30 Squadron at Sulaimania, with another Flight from 30 Squadron and a Flight of 55 Squadron at Kirkuk, was formed on 1 November, commanded by Squadron Leader R.V. Goddard*; its prime purpose at this stage being purely reconnaissance in support of the Iraqi army columns locally.

On 4 November, however, a rebel force led by one of Mahmud's adherents attacked and occupied the town of Penjwin, besieging its police post. That same day RAF aircraft

* Later Air Marshal Sir R. Victor Goddard, KCB,CBE,DL.

A constant stream of visiting civil aircraft, of many countries, became a familiar sight for all RAF stations in the Middle East in the years 1918-38. This De Havilland 60G Gipsy Moth was built as a single-seater for W L Hope of Air Taxis Ltd for his return flight to Kisumu, September 1928 and is seen here at RAF Ismailia. This Moth later achieved fame as Amy Johnson's '*JASON*' for her solo flight England-Australia 5/5/30 to 24/5/30, and is now preserved in the Kensington Science Museum.

EMBRYOS. Student pilots of No.4 FTS, Abu Sueir, Egypt practise formation flying in Avro 504J Lynx trainers.

bombed rebel picquet posts on hills over-looking the police post, causing casualties, and this immediate reaction so impressed the tribal leader Hama Rachid of Waina – formerly a follower of Mahmud – that he overwhelmed the rebel picquets and expelled all rebels from the town.

Mahmud by now was encamped in the hills north of Bayanan, moving from village to village exacting dues and recruiting men for his rebellion. Since nearly all such villages were professed pro-government communities, the RAF commander was reluctant to accede to government requests for such villages harbouring Mahmud to be bombed. Instead he arranged for air-dropped notices to all villages, declaring that any one found sheltering Mahmud might be bombed without further warning.

The onset of winter rains created a pause in activities but on 29 November rebel forces invested the village of Surdash, some 36 miles north-west of Sulaimania, and an army relief column, aided by 30 Squadron, marched and fought its way to the beleaguered village by 4 December, being rescued from a particularly dangerous ambush position en route by prompt air action. By this date the few RAF aircraft involved had already accumulated a total of 1,500 flying hours on recce and occasional bombing sorties.

Mahmud's defeat at Surdash kept him quiet until 3 January 1931 when he reappeared with a band of some 250 men in the Muan-Serao area, apparently intent on attacking the village of Halebja, but actually capturing the police post at Khurmal, then next day surrounding and attacking another police post at Shah Nadri. Despite appalling weather conditions hampering all ground movements, a relief force set out but on reaching Shah Nadri found that the rebels had, as usual, withdrawn, taking their looted arms and ammunition with them. Believing Mahmud still remained in the Halebja Plain, it was decided to 'sweep' the plain by ground forces, supported by aircraft from 30 and 55 Squadrons in a recce role, then once the rebels had been forced to shelter in villages, the RAF could undertake bombing operations to drive them into the open. This sweep operation commenced on 1 February but failed in its purpose when Mahmud and his forces escaped through mountain tracks and narrow passes north-east of the plain.

Once recuperated, Mahmud's next moves came in the first week of March 1931 when his men captured a convoy near Muan which provided the rebels with yet more looted arms and supplies, then he began moving towards the Qara Dagh, where Mahmud owned considerable property. On the way to Qara Dagh he and his force were caught in some watercourses south of Muan by aircraft from 55 Squadron which immediately attacked the rebels, while on 9 March another air patrol found a rebel force in the open near Naoti in the Qara Dagh valley and attacked it.

Until now all actions against Mahmud and his followers had achieved only minor successes, due partly to the reluctance of the RAF to bomb villages suspected of harbouring rebels, but by late March 1931 Mahmud's influence throughout southern tribes in Kurdistan was increasing greatly, threatening to precipitate a general revolt on a large scale. By then the RAF element had been joined by a Flight of 84 Squadron from Hinaidi and another Flight of 55 Squadron, and notices were again dropped on all villages concerned warning of air bombardment at only 30 minutes' notice should they shelter Mahmud's rebels; this warning was implemented against three villages on 28 March once the inhabitants had been seen to evacuate. By April Mahmud was located in the Qara Dagh range where the local Jaf tribesmen were in sympathy with his cause. Action came to a head on 5 April (Easter Sunday) when aircraft came upon a rebel force of about 100 horsemen south of the village of Awa Barika. Attacking these at once and reporting their position back to base, the leading pilot, Squadron Leader G.C. Gardiner* remained over the rebels until reinforced by all aircraft available from 30 and 55 Squadrons which spent the rest of that day attacking, refuelling, re-arming, and returning to the affray. These attacks inflicted severe casualties but Mahmud managed to escape in the hours of darkness via a solitary unguarded

* Later Wing Commander, DSO,DFC.

mountain pass over the Qara Dagh.

This engagement at Awa Barika proved to be crucial, in that Mahmud had barely escaped with his life, but had also suffered a severe setback with heavy casualties to his force. For the next few weeks he remained on the run, but his influence over the tribes had now been weakened considerably. Air actions continued daily, seeking out and attacking any rebel force located, including in April three Victorias from 70 Squadron and four Wapitis of 84 Squadron, supporting 55 Squadron. Failing to gain any more support from the Pizdher tribes or others, Mahmud finally struck south across the border and arrived in Piran, Persia on 23 April.

On 24 April Mahmud wrote a letter to Captain V. Holt, Oriental Secretary to the High Commissioner, revealing his desire to discuss terms of surrender. The eventual response on 2 May was an ultimatum offering his life to be spared, the grant of an allowance from the income of his estates, and permission to live with his family at some place selected by the Iraqi government, probably on the Euphrates. He was given until sunset on 13 May to decide, and on expiry of this limit Mahmud surrendered at Penjwin and was escorted to Sulaimania on the 14th, whence he was flown next day in a Victoria aircraft to Ur in southern Iraq and accommodated temporarily in the Railway Rest House. It was the end of Mahmud's sixteen years of being a constant thorn in the side of British and Iraqi authorities. Even in defeat Mahmud retained the reverence of his Kurdish tribesmen who, as he rode to captivity after his surrender at Penjwin, 'streamed down from the villages on the hill-sides to cluster round him and to kiss his hand, and the eyes of many were filled with tears as they bid him farewell' (*sic*).

Characteristically, a form of traditional British respect for a worthy foe was also accorded to Mahmud by his RAF opponents, who had long 'honoured' him with the unofficial title of 'C-in-C, RAF Training' in deference to the many years of operational practice he and his followers had provided for numerous air crews. Condemned by some as merely a tyrant, cruel and self-ambitious, perhaps a truer summary of Sheikh

The final surrender of Sheikh Mahmud, seen here (photo 50) with Iraqi and British personnel, including (photo 51) the RAF armoured car escort.

(*Left*) Station sign-post of RAF Habbaniya, with 'useful' information board saying 'London – 3287 miles' & 'Baghdad – 55 miles'.

(*Below*) DH9As of 'B' Flt, 47 Squadron from Helwan. Nearest is J7086 piloted by Flt Lt S D Macdonald (later, AVM,CB,CBE,DFC) whose gunner is holding Macdonald's pet dog 'Monty'.

Mahmud's character and aims was that given by Captain V. Holt, who was well acquainted with the sheikh and had been the officer specially selected to meet Mahmud at Penjwin on the occasion of his ultimate surrender: 'His greatest fault is that he was born a century too late.'

In that last period of anti-Mahmud operations, lasting from October 1930 to May 1931, the few RAF crews involved had mounted large totals of operational flying hours, as follows:

30 Squadron	2204
55 Squadron	1379
84 Squadron (One Flight only)	323½
70 Squadron	1142
	5068½

Mahmud's surrender had an immediate effect on various rebel sheikhs in sympathy with his cause. Karim Beg, a principal leader of the Jafs, made a dramatic appearance in Baghdad and pleaded for King Feisal's pardon on 27 May, while Mahmud Khan Dizli, a noted Persian outlaw, surrendered himself to Iraqi government authorities on 31 May. Other minor rebel leaders sought peace terms with Iraqi officials, promising to remain loyal to Feisal. In June Feisal and the High Commissioner made a tour of southern Kurdistan, meeting all the important tribal leaders and obtaining their assurances of future fidelity. The prime fact which emerged from all such meetings was the respect the tribes now held for the RAF's capability to react swiftly to any open revolt or rebel action. Prior to the advent of the aeroplane tribesmen had relied upon the inevitably slow response from any ground forces despatched against them, but aircraft had demonstrated conclusively that reaction would be a matter of mere hours, rather than days or even weeks by army or police expeditions.

While Sheikh Mahmud's surrender had concluded a particular lengthy campaign of operations, there remained in Iraq and its neighbouring countries a number of dissident tribes who resented both British and Iraqi government control. Chief among these in 1931-32 was Sheikh Ahmad of Barzan, whose personal authority extended over a vast area north-east of the Greater Zab river which stretched north-east to the Turkish border, within which the whole tribal population was hostile to any foreign administration. This area consisted of a mass of high mountains intersected by deep valleys and river gorges, with the lower slopes well-covered with scrub oak affording excellent shelter from any air or ground observation. The Iraqi government decided that the only solution to Ahmad's activities was complete occupation of the Barzan area by stages, using the Iraq army to establish a series of armed posts at strategic points to retain permanent control.

Initially, the RAF was merely to supply air co-operation for the Iraq army columns, and a Flight of No 30 Squadron was detached to the desert landing strip at Diana, some 80 miles north-east of Mosul, while the rest of the squadron remained at Mosul in reserve. At Diana, 30 Squadron's detached Flight was joined by elements of the embryo Iraq Air Force*, and these were to be air-supplied by Victorias of 70 Squadron.

The Iraq army advance commenced on 15 March 1932, having to fight tribal opposition intermittently, but on 3 April one column was attacked by a large force of rebels near Wazi and was only saved from annihilation by the intervention of aircraft from Diana which bombed and strafed the opposing tribesmen continuously throughout the hours of daylight, losing one pilot who died of wounds (Sergeant H.V. Hudson) with his gunner, AC1 T.F. Merrick wounded. Despite mortal chest wounds Sergeant Hudson had accomplished a safe landing at Diana, thereby saving his gunner's life, but succumbed to his own injuries shortly after. Most of the other aircraft involved returned with plentiful bullet holes to testify to the fierce, accurate ground fire encountered.

This army column was slowly withdrawn in the face of

* Formed with a nucleus of five DH Gipsy Moths, fitted with bomb-racks, wireless, cameras and extra fuel tanks, in 1931 and manned by Iraqi pilots trained at Cranwell. In April 1932 these were reinforced by three DH Puss Moths, similarly armed for bombing etc.

Short Rangoon flying boat, S1433, seen here on test prior to delivery to No.203 Squadron at Basra, Iraq. Later civilianised as G-AEIM with Imperial Airways, this particular Rangoon was finally scrapped in 1938.

Close-up of the mid-fuselage gunners' cockpits of a 203 Squadron Short Rangoon.

continuing opposition, and the RAF resumed active operations, strafing rebels caught in open country, while the Victoria transport aircraft dropped much-needed supplies to the ground troops. This setback caused the Iraq authorities to declare that the army could do no more alone, and formally request the RAF to take over control of operations. This was agreed, and while a suitable air element was assembled, a Victoria aircraft flew over all important rebel villages announcing that air action would begin from 28 April by verbal warnings in Kurdish dialect delivered via loudspeakers fitted under the Victoria's fuselage while flying at 2/3000 feet; a 'voice from the sky' which made a great impression on all villagers.

The new air 'striking force' (*sic*) consisted of the whole of No 55 Squadron, most of 30 Squadron, and a Flight of 70 Squadron, all being based at Erbil, 80 miles east of Mosul, from 26 April, while other aircraft still based at Diana would help when not needed for army co-operation duties with the Iraqi troops. On 26 April, however, a recce aircraft force-landed near Shirwan a Masin, its pilot sustaining a broken shoulder, and its crew became prisoners of Sheikh Ahmad who requested a doctor to attend to the pilot. This was used as an opportunity to persuade Ahmad to agree to terms, and negotiations were continued until 22 May. With no agreement reached, air operations resumed on 25 May when villages – previously warned and therefore evacuated in good time by their inhabitants – were duly bombed with light practice smoke bombs only with a few delayed action explosive bombs scattered around the villages' outskirts to discourage re-occupation. The rebels retired to nearby mountain caves but these were then attacked continuously for six days, causing Ahmad to change his location almost daily and eventually to be driven to the Turkish border, which many rebels, their families and flocks crossed in early June.

Now out of touch with his main rebel forces, Ahmad negotiated with the local Turkish commander and on 21 June crossed into Turkish territory and surrendered himself to the senior Turkish authority. The Iraqi army now moved quickly and occupied Ahmad's former lands and civil administration

The ground crews' problems came in all shapes, exemplified here by DH9A, E9918 of 55 Squadron at Hinaidi, ca, 1929-30 which tried a 'one-point' landing …

Fairey IIIFs of 45 Squadron, based at Helwan, Egypt ca. 1930.

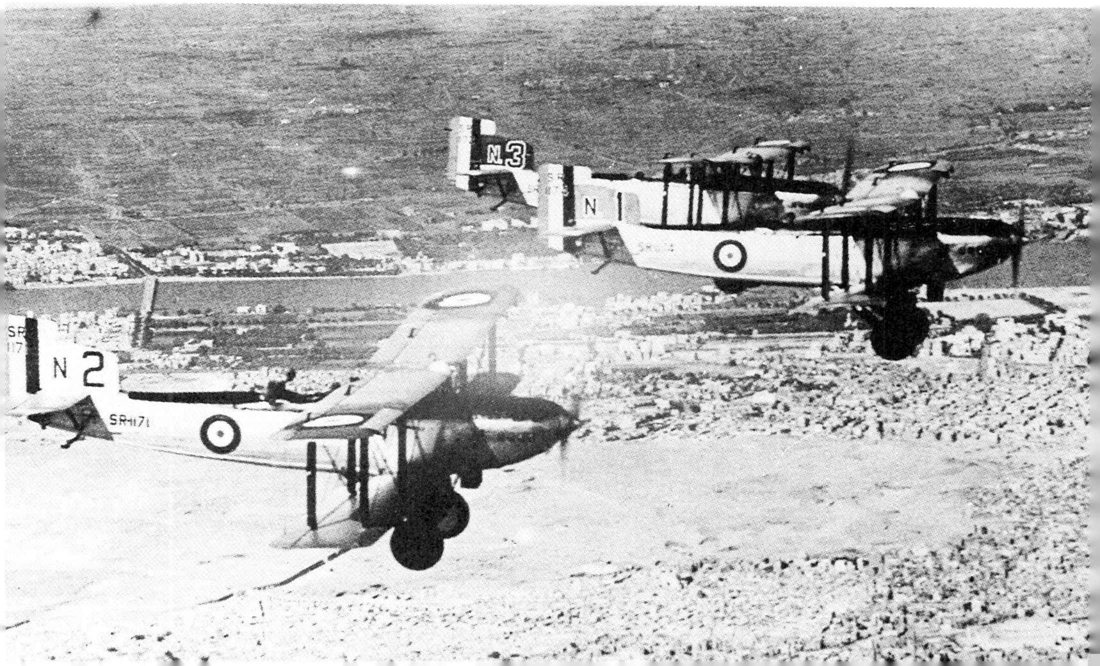

Fairey Gordon, JR9806 of No.6 Squadron. The 'R' added to the serial prefix indicated a repaired/rebuilt aircraft.

Hawker Hardys of No.30 Squadron at Mosul, marked with the unit badge – a palm tree – on fins. 30 Squadron was the first to receive Hardy aircraft, re-equipping from Wapitis in April-August 1935 at Mosul.

was established. While air action had undoubtedly saved part of the Iraq army from decimation at one stage of these Ahmad operations, the prime role of the RAF remained simply air reconnaissance and supply for the ground forces. By the terms of an Anglo-Iraqi Treaty in June 1930, and bearing in mind the imminent full independence of Iraq due in September 1932, the RAF had agreed to withdraw from Hinaidi and Mosul within five years of Iraq's independence date, and would instead maintain a reduced force at a new RAF base to be constructed at Dhibban, roughly 50 miles west of Baghdad, near Lake Habbaniya.*

At the moment that Iraq assumed total independence the RAF in that country comprised Nos 30, 55, 70 and 84 Squadrons, apart from various armoured car sections, supply and aircraft depots. No 8 Squadron had left Iraq in February 1927 to move to Khormaksar, Aden where, with a break of four years (1942-46), it was destined to remain based until the ultimate British withdrawal from Aden in 1967. Nos 30, 55 and 84 Squadron continued to provide the troop and supply transport 'Carter Paterson' (*sic*) services with its lumbering Victorias. Further west, in Transjordan, was No 14 Squadron based at Amman and equipped with Fairey IIIFs from December 1929 which were replaced by Fairey Gordons in late 1932.

From 1932 full responsibility for administration, security and overall legal control of Iraq was in the hands of Iraq's own government, and the continuing smaller RAF presence was primarily in the interests of protection for British residents and properties, although for the next few years until the outbreak of World War Two the RAF squadron *could* be called upon to assist Iraqi authorities in maintaining law and order provided formal prior permission had been sought, and granted, from the UK government.

One fresh unit to the Iraqi scene was No 203 Squadron. Reformed at Plymouth on 1 January 1929, equipped with

* Later retitled RAF Habbaniya on 1 May 1938 and base for No 4 Flying Training School (FTS) in World War Two. The base was ultimately handed over to the Iraqi government on 2 May 1955.

Supermarine Southampton flying boats, the unit left the UK on 28 February that year and eventually arrived at Basra, its new base, on 14 March, where its main role was simply to recce the area's coastal waters and generally 'show the flag' to the local tribal inhabitants. From Easter 1931 the squadron exchanged its Southamptons for Short Rangoon flying boats* which continued the unit's roles, including patrolling the Persian Gulf in co-operation with Royal Navy vessels in thwarting drugs and guns' smuggling, and transporting various industrial teams and equipment seeking oil sites.

The future main RAF base in Iraq, Dhibban, was steadily built and extended during the years 1934-36, and then began receiving the various units scheduled to be resident there. From October to December 1936 No 30 Squadron, the RAF Supplies Depot, and No 1 Section, No 1 Armoured Car Company joined the RAF Depot from Hinaidi in moving to Dhibban, while in 1937 Dhibban SHQ was officially established from 1 January, followed by the transfer of No 55 Squadron in September, and No 70 Squadron in October. Nos 84 and 302 Squadrons remained based at Shaibah and Basra respectively.

Hinaidi, formerly the central RAF base in Iraq, now became the premier airfield and depot for the Iraqi Air Force, which had steadily acquired numbers of converted De Havilland civil aircraft, e.g. eight DH 84M 'Dragon' twin-engined aircraft in 1933 which had been modified to include bomb racks, rear gunner's cockpit, and a two-gun fixed installation in nose sections. Piloted by RAF-trained Iraqis, with a sprinkling of regular RAF personnel on loan for overall guidance and supervision, these Iraqi Air Force units now assumed responsibility for quelling or preventing any tribal disturbances and generally providing air co-operation for the Iraqi Army and its associated Levies' formations.

The five RAF squadrons which were to remain in Iraq until the outbreak of the 1939-45 war spent the later 1930s in routine duties, and in 1935 four of these began receiving new

* Only six Rangoons were ever built for the RAF and, after 203 Squadron exchanged these for Short Singapores in 1935, all went to 210 Squadron in the UK.

Vickers Vincent of 55 Squadron over Iraq. 55 re-equipped from Wapitis in January-March 1937.

aircraft designs. In January 84 Squadron replaced its Wapitis with Vickers Vincents; No 30 Squadron began converting to Hawker Hardys in April; 203 Squadron started exchanging its Rangoons for Short Singapore IIIs in September; while 70 Squadron 'pensioned off' its ancient Victorias for a stablemate design, the Vickers Valentia. Finally, in February 1936, No 55 Squadron began receiving Vincents in place of its well-worn Wapitis. All these aircraft designs were still out-dated biplanes, and the only more modern type to reach an Iraq-based RAF unit before 1939 was the Blenheim I bomber; the first example of which reached No 30 Squadron at Dhibban on 13 January 1938, beginning this unit's full conversion to a 12-aircraft establishment over the following months.

To most of the several thousands of RAF men who served in Iraq during the years 1918-38, particularly in the 1920s, it had been an unloved, unwanted posting; one generally considered on the RAF 'grapevine' as possibly the worst posting of their careers in terms of living and working conditions, though it must be said that in slightly later years Aden assumed priority in that context. Yet Iraq had been the first testing ground for Hugh Trenchard's policy of 'air control' and had, under John Salmond and his succeeding commanders, proved that policy to be wholly effective, and enabled the scheme to be applied to other British mandated territories. That success in Iraq might well be summarised by the first remarks made by Sheikh Mahmud when he finally surrendered at Penjwin in 1931. Going up to the nearest uniformed RAF officer, the sheikh tapped the RAF wings on the officer's tunic and said solemnly, 'You are the people who have broken my spirit ... but I taught you to bomb.'

CHAPTER SIX

Aden and Palestine

In view of the success of Trenchard's 'air control' in Iraq, this scheme was then extended to other British mandates and protectorates, including Aden and Palestine. Aden, some 100 miles east of the southern entrance to the Red Sea, on the southern coast of the Arabian Peninsula, had first been annexed in 1839 by a naval and army expedition mounted by the East India Company, the commander of which force, a Captain Haines, Indian Navy, became the first British Resident in Aden.

In 1872 Turkey began extending the Ottoman empire by invading Yemen to the north of Aden and advanced southwards to the borders of Lahej, the Sultan of which area was bound by treaty to Britain. However, the Turks were persuaded to withdraw back into the Yemen after representations by the UK government, and Britain was able to gradually conclude numerous treaties with the various Sultans of lands adjacent to Aden. With the opening of the Suez Canal in 1869 Aden grew prosperous with its increased use by shipping to and from the Far East, but after the outbreak of the European war in 1914 Turkish forces in the Yemen tried again to oust the British from Aden, advancing eventually to occupy Lahej, then most of the Aden harbour area, including Little Aden which had been a British possession (purchased in 1888) and the town of Sheikh Othman, also acquired by the British by purchase in 1882.

The first military aircraft to appear over Aden were Short seaplanes carried aboard HMS *Raven II* in 1916 which patrolled over the Turkish positions, dropping leaflets to entice local Arabs to revolt against Turkish occupation, and carrying out occasional bombing attacks against Turkish centres such as

Lahej. The first air unit to be actually based on Aden was the Aden Flight, formed at Lahore, India from Henry Farman aircraft and personnel provided by Nos 31 and its off-shoot 114 Squadrons. This Flight operated against the Turks from 1917 until the Armistice and then returned to India in early 1919. For the next few years small detachments of RAF and Fleet Air Arm units operated spasmodically from Aden against tribes threatening peace in the surrounding protectorate territories, including punitive bombing raids against dissidents in British Somaliland.

Examples of such local use of air power included two Sopwith Snipes demonstrating over a Yemeni town where a British colonel had been made captive in late 1919, resulting in the officer's immediate release; while in 1922-24 a pair of Bristol F2Bs carried out bombing raids against rebelling tribes in Somaliland after a local Political Agent had been murdered. This pattern of maintaining a few RAF aircraft at Aden primarily as a safeguard for friendly Sultans' lands against inter-tribal activities continued until 1927, when it was decided to base a complete squadron at Aden. The unit selected was No 8 Squadron, based at Hinaidi, Iraq, commanded by Squadron Leader W.A. McClaughry, DSO,MC,DFC,* equipped with DH9As which moved en bloc to Aden in February 1927.

No 8 Squadron had no sooner settled in to its new base at Khormaksar, Aden than it was called upon for operations. Yemeni forces invaded Aden protectorate country and seized the town of Dhala in late 1927, and in February 1928 a force of Zeidi tribesmen invaded the Koteibi and Alawai districts and kidnapped two sheikhs who had treaty relations with the British. At the beginning of 1928 Aden had become another area to be placed under the overall command of an RAF officer, in this case Group Captain W.G.S. Mitchell, CBE,DSO, MC,AFC (later ACM Sir William, KCB,CBE,DSO,MC,AFC).

After the usual 48-hours' notice pre-warnings of air action had been dropped on Yemeni positions, eleven Ninaks of 8 Squadron commenced attacks on Kataba on 21 February and

* Later AVM W.A. McClaughry, CB,DSO,MC,DFC and killed in an aircraft accident 4/1/43.

continued these sorties for two more days. Pausing to ascertain any effect from these raids, and on failure of the kidnappers to release the two sheikhs, air action was resumed on 10 March against Kataba, Mafalis and the surrounding countryside where large troop concentrations had been reported.

This action continued until 25 March when a truce was implemented at the request of the Imam of Yemen who released his captives on that date and asked for a meeting with government representatives. The truce extended until 7 July on condition that Yemeni forces evacuated Dhala by 20 June, but the Imam merely used the time to recruit more men throughout southern Yemen with the intention of raiding Lahej, and meantime increased his garrison at Dhala. Accordingly, air action was resumed again from 25 June, bombing the towns of Taiz, Kataba, Mafalis, Yerim and the Dhala Fort. This air offensive gave encouragement to several 'friendly' tribes who proceeded to engage the Zeidi invaders and, in protracted clashes until late August, finally drove out the Zeidis from all their 'occupied' centres.

The operational efforts by No 8 Squadron during these actions were not small. In the first phase, February-March, the DH9A crews had dropped over 40,000lb of high explosive bombs and 1200 incendiary bombs during 332 operational flying hours; while in the period June-August they had flown 927 hours, dropping 49½ tons of bombs and firing some 33,000 rounds of ammunition. As Trenchard reported to the Cabinet (somewhat smugly), 'These operations were successfully carried out in less than six months by one squadron at a total cost of £8,567, with only one British casualty.' Still engaged daily in fighting for the autonomy of 'his' RAF, Trenchard rarely overlooked any opportunity to press his claims for wider and economic use of the Service to protect Britain's overseas responsibilities.

As an RAF base, Khormaksar proved to be worse than anything No 8 Squadron's personnel had experienced in Iraq. As one visiting RAF officer* recalled the place:

* The late Air Commodore the Hon E.L. Howard-Williams, MC,MLC,MA,DIC.

R. O. T. B.

Rock of Aden, Woe is me;
Let me take myself from thee,
Let me count the days to go;
Till I may my kit-bags stow,
Then may I thy beauty note,
From the "blunt end" of the boat.

D. T. C.

ROLL ON THAT BOAT. The constant cry of the airmen serving overseas ...

> I remember Aden as a dyed-in-the-wool dump in the desert. The landing ground – they were never called airstrips or aerodromes then – was just a patch of sand. The squadron's quarters were rough and ready. Heavens, but it was hot! There was no air-conditioning, little ice, enough water (most of it boiled), and ceiling fans raced everywhere like huge daddy long-legs.

Aden's 'summer season', May to September, produced a humidity of 80 per cent-plus with temperatures of 100 degrees F or more, while the 'winter' months proved little more pleasant. A mere four to five miles eastwards, across the harbour waters, lowered the infamous Barren Rocks of Aden; a mountainous range stretching inland which pressed all habitation of the port into a thin, featureless coastal strip of sand and rock-dust. Moreoever, take-off from Khormaksar in virtually any direction meant flying over water within seconds, with an inevitable ducking should any aircraft engine falter as it strived to attain a safe altitude. As the late Sergeant Eric Wilkinson put it:*

> I'd seen service in Egypt, Iraq and India, but Aden was undoubtedly the bottom of the pile as far as postings went. In the late 1920s when I arrived there, Khormaksar was something of a Hell-hole as far as living and working conditions were concerned. Servicing our aircraft out in the open sunlight, sweating like pigs non-stop, then returning to our billets which had nothing but damned great wooden fans to cool us off – all they did really was circulate the hot air and stir up the dust – was no picnic.

With characteristic Service adaptability, the men of 8 Squadron proceeded to steadily improve their accommodation, both domestic and technical, and settled down to a routine of regular reconnaissance patrols – mainly mapping the surrounding desert lands, but also keeping a wary eye on the nearer Yemeni towns and other Arab villages – and maintained

* Correspondence with author, 1973

at least one Flight at firstline 'readiness' state in case of any emergency requiring actual operations. In the event, apart from a lengthy detachment to British Somaliland in 1935-36, No 8 Squadron was destined to remain based at Khormaksar until 1942, then returned to Aden in 1946 and saw two more decades of faithful service at Khormaksar until the British evacuation of the colony in November 1967.

In 1928 the unit began phasing out its well-worn DH9As as it received Fairey IIIF aircraft replacements, these latter being of near-whole metal construction with 570hp Napier Lion engines. Though more comfortable to fly, the IIIFs offered only slight improvements in speed, range and bomb load capacity. With its new aircraft 8 Squadron was soon involved in active operations again in early 1929. One large tribe, the Suhebis, had for some years been in receipt of a system of payments as a guarantee of good behaviour, but when these payments ceased the tribes raided the Lahej district, stealing cattle and destroying property. Bombing sorties commenced on 30 January and continued until 5 March, mainly using small BIB incendiaries to destroy crops and dwellings. By 27 March all rebel chiefs had come to terms and paid compensation fines.

In the early 1930s the prime task of British ground and air forces based in Aden was improvement of communications between the myriad villages and tribal towns, mainly in the context of roads but also by the establishment of dozens of temporary landing grounds for aircraft throughout British-protected areas. While providing air support for all such ground expeditions, No 8 Squadron was to be called on constantly to provide aerial demonstration flights over potential tribal trouble-spots; such flights, allied with air-dropped leaflets warning of possible air raids, usually resulted in would-be rebels coming to terms without any need for air operations, such was the contemporary respect for potential air action held by most tribesmen. Other more mundane duties for 8 Squadron included a regular fortnightly delivery of mail by air between all salient centres, and constant provision of air transpor-

Vickers Vincent, K4129 of No. 8 Squadron, Khormaksar, Aden.

Fairey IIIFs of No.8 Squadron over the 'Barren Rocks of Aden', from their base at Khormaksar, Aden.

tation for political officers and other officials for visiting outlying towns and villages.

These routine matters were only occasionally interspersed with actual bombing sorties; more usually a demonstration flight by one or two aircraft over any troublespot was sufficient to bring the dissidents to peaceful terms. Even when bombing was carried out – and then only as a last resort – ample pre-warning was given to the tribes, enabling them to move their families and flocks away from the target villages, thereby keeping human casualties to an absolute minimum (if indeed any). It was a procedure well understood by all tribes and accepted with little rancour in a majority of cases. Inevitably, by the nature of the tribes, many such incidents necessitating the involvement to some degree of the squadron were simply intertribal feuds and traditional hostilities whose roots were buried in the mists of history, leaving the RAF to simply adopt a patient police role in preserving law and order generally.

This overall pattern of duties for No 8 Squadron was to remain the norm throughout the 1930s, the premier source of unrest being the constant threat of invasions by sections of Yemeni tribes across the Yemen border. In early 1935 the squadron again exchanged its aircraft, receiving Vickers Vincents in place of its Fairey IIIs, which it was to continue flying until 1942. In 1935 too, the squadron received unexpected reinforcements when Nos 12, 41, and 203 Squadrons were detached to Aden. This sudden strengthening had resulted from the failure of the League of Nations to prevent an Italian invasion of Abyssinia. With the possible need for protection of British interests in the Sudan, Kenya and British Somaliland, the British government mobilised a number of UK-based (mainly) squadrons to transfer to the Middle East areas.

The units detailed to move to Aden for this crisis were No 203 Squadron from Basra, flying Short Singapore III flying boats, commanded by Group Captain R.E. Saul, DFC (later Air Vice Marshal); No 12 Squadron from Andover, flying Hawker Harts, commanded by Squadron Leader A.C.

Bayley; and No 41 Squadron from Northolt, flying Hawker Demon fighters and commanded by Squadron Leader J.A. Boret, MC,AFC. All arrived at Aden in October 1935 and were to remain there until 1936; 203 Squadron leaving in July, while Nos 12 and 41 Squadrons commenced repatriation to England from 11 August 1936. In the interim all three units participated with 8 Squadron in overall patrol and other operational duties.

Apart from a handful of detached RAF and Fleet Air Arm aircraft, No 41 Squadron was the first-ever fighter unit to be based at Aden. By early 1937 No 8 Squadron had resumed its solitary position as 'Aden's Own' RAF unit, and on 1 April that year Aden became legally a Crown Colony. This change in status for the protectorate was primarily an administrative move, exchanging the previous 89 years of (theoretical) administrative control of Aden by the Indian government for future responsibility by the UK Colonial Office. Since India had never provided any finance for development or welfare in Aden, and the Arab population had never liked the link with India anyway, the change was generally accepted with approval.

Unlike Iraq, the position regarding Britain's post-1918 given mandate of Palestine, despite eventual responsibility for control of the country passing to the RAF, was an entirely different matter in the context of air control. Britain and her allies had conquered Palestine in 1918, defeating the Turkish rulers who had controlled the Holy Land for centuries. While still at war the British government had sought support for its cause from both the Arab population and the Jewish Zionists, promising both to support their own nationalistic ambitions once the war had been won. When peace did arrive, however, Britain, for a variety of diplomatic, strategic and political reasons, leaned towards the Arab cause and, in effect, gave Arabs an inferred justification for all future attempts by the Arab nations to drive out of Palestine the many Jewish settlers attempting to recreate their ancient homeland as a Jewish State in its own right.

From the RAF's viewpoint any control or security of Palestine was inexorably tied in with its neighbouring British mandate Transjordan – indeed, both were looked upon initially as simply one mandated territory under the aegis of a single High Commissioner based in Jerusalem. Transjordan, with its common border to Ibn Saud's Arab Nejd country, was in effect a bulwark against any attempted Arab incursions into Palestine from the Nejd area. Nevertheless, Palestine itself presented a different problem for any form of control to such mandates as Iraq, in that it was not an undeveloped country even in the early 1920s, and therefore any internal disorders or security problems would mainly be the responsibility of internal police and military forces. In summary, although an RAF responsibility notionally,* full air control methods in Palestine and Transjordan were never really applicable owing to the nature of these particular mandates.

Although, as stated, internal law and order within Palestine and Transjordan was chiefly the responsibility of ground forces, civil and military, such as the Arab Legion and police services, the need for an RAF presence was considered vital, if only as a big stick for final resort in any serious situation. The premier RAF unit to serve in these mandates throughout the 1920s and 1930s was No 14 Squadron RAF which had served in the area during World War One. First disbanded in England on 4 February 1919, the squadron was reborn at Ramleh, Palestine on 1 February 1920 by the process of renumbering No 111 Squadron there. Equipped with Bristol F2B aircraft inherited from 111 Squadron, the unit was designated for army co-operation duties.

It soon became used on occasion to assist local security forces in quelling tribal disturbances, in close liaison with RAF armoured car sections; an example occurring in September 1923 when the armoured cars, closely supported by 14 Squadron's F2Bs, engaged a large force of mounted

* Palestine and Transjordan were placed under the command of an RAF officer for operations from February 1924.

tribesmen loyal to Sultan Adwhan which had captured several Arab Legion outposts, cut telephone and telegraph lines, and blockaded traffic using the Amman-Jerusalem road. The armoured car crews were successful in dispersing this force, aided by forward reconnaissance by 14 Squadron, resulting in the rebellious Sultan fleeing to Syria.

In 1924 the squadron began receiving a few DH9As to supplement its Bristol F2Bs – the Ninaks having a greater range for distant patrols and/or operations – and the 'new' aircraft were quickly to be engaged in active operations. On 14 August 1924 a force of Wahabi Arabs began raiding in the Ziza and Yadudi districts, some twelve miles south of Amman. Within the hour of receiving reports of these raids, three DH9As took off and attacked the raiders with bombs and machine guns, along with three RAF armoured cars. The DH9As, joined by two Bristol F2Bs slightly later, refuelled and re-armed, then continued their assaults on the Wahabi force – estimated at some 3-4000 men – and killed an estimated toll of at least 500; the only RAF casualties being three men slightly wounded. The Wahabis became a 'disorganised rabble' (*sic*) and fled 'with all speed', while several Wahabis made prisoner stated that they did not mind rifle fire and would readily have attacked any ground forces, but were terrified of bombs from the air, which they did not understand. The Palestine/Transjordan Administration Report for 1924 when referring to this successful operation went so far as to state 'the prompt and energetic action of the RAF averted a serious menace to the peace of Transjordan and the safety of its capital.'

In January 1925 No 14 Squadron detached a Flight to Beersheba to co-operate with a composite force of Palestine police and the Arab Legion for reconnaissance and assistance in the arrest of certain tribal outlaws. The aircraft provided forward recce for the ground elements and also supplied these with rations, water and fodder, while RAF tradesmen operated a W/T pack set mounted on two camels for 'ground liaison' – the first such successful use of camel-borne W/T in the country.

This was a typical small example of the type of overall air co-operation with the civil and military security forces in

DH9A, ER959 of 14 Squadron at Amman, Transjordan.

DH9A, JR7827 of 14 Squadron over the Jordan valley, 1929.

DH9A, ER927, 'B3' of 14 Squadron over scrubland at El Gueira, Transjordan – uninviting territory in the event of a forced landing

Palestine and Transjordan which 14 Squadron was to offer over the following years. In many such instances the actions taken were preventative rather than a cure, as in the spring of 1926 when, while French forces were operating against the Druze revolt in Syria, large numbers of Druze refugees crossed the frontier into Transjordan and were permitted to set up a camp at Kasr Azrak on the condition that this would not become a resort for armed Druze men.

By late 1926 it became apparent that Azrak was becoming a safe retreat for armed Druze fighters and it was decided to enforce the neutrality of the area. Accordingly a force comprised of three sections of No 2 Armoured Car Company, RAF and three companies of the Transjordan Frontier Force, collectively titled 'Azrakcol' and commanded by Group Captain Lionel W.B. Rees, VC,MC, moved to and occupied Azrak on 17 April 1927 where martial law was proclaimed and all known Druze men classified as 'undesirables' were rounded up, court-martialled and sentenced. By 10 May the potential crisis had been completely quashed. As ever, No 14 Squadron had provided constant air support and reconnaissance for the ground forces, while Azrakcol HQ staff had been transported from Amman to Azrak by Victorias of No 216 Squadron detached from their normal base at Heliopolis, Egypt. Again, the official summary was moved to comment, 'the situation generally was controlled in a manner that was only made possible by the co-operation of aircraft.'

By 1926 No 14 Squadron had relinquished the last of its Bristol F2Bs and was now fully equipped with DH9As, and in the summer of that year the unit moved base to Amman in Transjordan, the Air HQ for the dual-mandated territories. The squadron was to remain based at Amman until August 1939, where it was in a better location for assisting prevention of the spasmodic raids into Transjordan areas from across the Syrian border by such tribes as the Arab el Jebel.

A typical example of this tribe's depredations occurred in March 1929 when sections of the Arab el Jebel poured across the Syrian boundary to plunder the flocks of the Beni Sukhr tribe. Eight DH9As from 14 Squadron, supporting a section of

No 2 Armoured Car Company, RAF and troops of the Frontier Force, left Amman on 13 March and eventually caught up with the raiders near Mafrak. Aircraft and ground troops immediately attacked, causing casualties and forcing the tribesmen to abandon most of their looted stock as they fled across the Syrian border. With strict orders to take action only over Transjordan territory, the DH9A and armoured car crews had to cease their attacks at the border. It was but one of the many such operations undertaken by the RAF throughout 1929 and early 1930, a period of virtually continuous reconnaissance and co-operation with ground forces for 14 Squadron and the detachment of Victoria transport aircraft of 216 Squadron, with occasional reinforcement from No 6 Squadron.

A number of Arab-Jew disturbances within Palestine which began reaching serious proportion in August 1929 instigated the move of AHQ, RAF Palestine and Transjordan from Amman to Jerusalem in that month. The steadily escalating Arab hostility towards 'excessive' (*sic*) Jewish immigration into Palestine began to reach boiling point in the early 1930s, exemplified in late 1933 when Arab protest demonstrations, despite police warnings of consequent action by the security forces, took place in and around Jerusalem and Jaffa. On 28 October nine aircraft from 14 Squadron demonstrated over such disturbed areas as Ramleh, Jaffa, Haifa, Nablus and Jerusalem, and over the ensuing weeks succeeded in dispersing several large gatherings of rebel Arabs merely by diving very low in mock strafing attacks without needing to fire or bomb. It was simply the latest of many Arab attempts to eject all Jewish people from Palestine which were to increase in intensity for many decades, and in which the RAF elements were to play a major part until 1948.

By 1936 this constant tension between Arab and Jewish communities within Palestine was brought to a head by a minor incident on 15 April, when a Jew was killed and two others injured by Arab bandits in the Tulkarm area. Though devoid of political motive, this incident proved to be a spark for the eruption of general and bloody disorders lasting, with rare and brief respite, for several years. Jewish reprisal actions led to more disturbances and on 21 April the Palestine Arab Party called for a general strike, which in turn led to a wave of murders, arson, sabotage and crops' destruction over a very widespread area.

The existing British garrison in Palestine was swiftly reinforced by land forces based in Egypt and Iraq, as was the RAF. No 14 Squadron which had exchanged its doughty DH9As for Fairey IIIFs in December 1929, then reconverted onto Fairey Gordon aircraft in September 1932, was the only complete squadron based in the area, but was soon to be reinforced by several other RAF units which had recently arrived in the Middle East zones, while No 6 Squadron had detached a Flight to support 14 Squadron, followed soon after by the remainder of 6 Squadron and a Flight of six Hawker Harts of No 33 Squadron initially, joined by the rest of the unit by the end of August 1936. No 33 had been moved to Mersa Matruh, Egypt from the UK in October 1935 as part of the RAF's strengthening of all RAF units in the Middle East consequent to the Abyssinian Crisis.

Further army reinforcements joined the Palestine garrison in June and all quickly became embroiled in the unceasing operations against not only politically-motivated Arab mobs attacking Jewish settlements and other properties but also non-political gangs of robbers and murderers seizing the opportunity to loot and plunder in the outlying urban areas; apart from pure fanatical terrorist groups – Arab and Jewish, political in the main – whose activities could never be predicted. Clashes with all of these various groups inevitably incurred casualties on both sides, and where aircraft were involved in support of the civil and army forces it became necessary to resort to light bombing and machine gunning on many occasions, being invariably greeted by ground fire which included occasional machine guns.

By September the increased determination of certain armed bands to press home their attacks demonstrated a level of careful planning by leaders who clearly had military experience, and this was confirmed by reports of the arrival of two rebel leaders on 1 September, Fauzi Al Qawaqji – an ex-Iraqi army officer – and Adil al Azmah from Syria; an obvious indication that many rebels were Syrians or Druze who were in control of armed gangs in the Tulkarm-Nablus-Jenin area, while dead rebels were found to be well-armed with plentiful ammunition.

The British response was to further increase reinforcement of the garrison forces, and from 15 September, to place an army officer as GOC all forces in place of the existing RAF officer, on which date Lieutenant-General J.G. Dill, CB,CMG,DSO took up the new appointment. The situation remained critical, with terrorists and other armed bands maintaining their activities during October, despite the cessation of the Arab general strike on 12 October. Hard fighting, involving the RAF, continued until November by which time the country became far less disturbed apart from isolated non-political incidents by outlaws, and No 33 Squadron was returned to Egypt along with two Flights of 208 Squadron.

These operations had not been free of casualties for the RAF. On 3 September aircraft from 6 Squadron had been called on to rescue a detachment of the Lincolnshire Regiment picketing the Nablus-Tulkarm road from a larger force of Arabs firing down on them from nearby higher ground. One Hawker Hart, crewed by Pilot Officer T.B. Hunter and AC1 E.G. Lincoln, was shot down, killing both men in the crash, while the unit commander, Squadron Leader H.M. Massey (later, Air Commodore, CBE,DSO,MC) was wounded but managed a safe landing at the nearby Tulkarm landing strip. The action lasted all day without pause, finally forcing the Arabs to retreat to hill refuges and then disperse. On return to base one airman gunner was astonished to discover five bullet holes or slashes in his overalls – none of the bullets having even scratched him!

The bitter fighting in 1936 also cost another 6 Squadron pilot, Sergeant Tarrant-Packer, whose aircraft was crippled by ground fire. He took to his parachute but was never heard of again. Decorations to 6 Squadron for its part in the 1936/1939 operations eventually trickled through the official channels, and amounted to four DSOs (including one to Squadron Leader E.R.C. Hobson, DFC,OC Squadron in 1938), eight DFCs, seven DFMs and thirteen 'Mentioned in Despatches'.

The year 1937 proved to be generally turbulent with occasional Arab-Jewish clashes and acts of pure terrorism on both sides, but not on the intense scale of the previous year until late December when a detachment of the Transjordan Frontier Force was ambushed by a force of some 200 Arabs. An emergency call for air support at Arraba was quickly answered and aircraft crews soon drove the Arabs into the hills, then continued harassing them until they had split into small groups in the Galilee district. This latest clash was merely a preliminary to 1938 when rebellion, terrorism, and direct conflict with government forces reached a new peak of intensity which was to last throughout the year.

As the year commenced many areas of the country, such as Galilee and Samaria, were infested with large rebel bands who not only attacked government forces but indulged in acts of terrorism in order to subjugate many outlying villages as rebel strongholds. The land defence forces in Palestine in January 1938 consisted of two infantry brigades, the Palestine Police Force, and some units of the Transjordan Frontier Force stationed in the Jordan Valley area. The RAF had only two Flights of No 6 Squadron (increased to three by February), No 2 Armoured Car Company, and across the border at Amman No 14 Squadron, which was about to convert from Fairey Gordon aircraft to the Vickers Wellesley monoplane bombers; the first batch of which began arriving in March, the squadron becoming fully converted by June.

January 1938 proved to be an active month for the men of 6 Squadron who flew totals of 57 ground co-operation sorties and 23 general reconnaissances over suspected rebel centres. In addition air action was found necessary on a few occasions, notably on January 31 when an army column was besieged by a force of at least 100 Arabs at Umm Al Fahm and immediately signalled for air reinforcement. Five of No 6 Squadron's crews

RAF station, Amman, 1929, base for No.14 Squadron.

Vickers Wellesleys of 14 Squadron over the Dead Sea, carrying their 'Munich' code letters BF and the official unit badge on the aircraft fins.

responded to this emergency and kept up bombing and strafing attacks until dusk, despite no small damage to the aircraft by return ground fire.

This pace of daily co-operation and recce flights, interspersed with occasional active operations, continued through February, during which month rebel bands began increasing in numerical strength, reflected in more daring acts of terrorism and confrontations with land forces and police. This trend continued in March and resulted in actions which gave the rebels their greatest defeat to date. On 4 March an armed band of some 400 rebel Arabs was engaged by land and RAF forces near Yamun in the Jenin district, while a second dissident force was also attacked in northern Galilee on 8-11 March; both clashes resulted in a heavy rebel casualty toll, including the killing of one particularly notorious rebel leader Sheikh Attiyeh.

Despite this clear setback, rebel leaders now proceeded to employ much smaller raiding bands over wider areas, resulting in a sharp increase in the number of murders and acts of sabotage and terrorism in April.

The chief political event of that month was the arrival in Palestine on 27 April of the Partition Technical Commission which then toured Palestine and Transjordan attempting to gather views and evidence, but met with an Arab boycott whereby Arabs refused to offer any evidence. This ill-fated commission eventually left the country on 3 August having achieved little of worth in resolving the Arab-Jew problem. Meanwhile No 6 Squadron and the RAF armoured cars were kept busy, not only with daily routine patrols and duties but active operations on several occasions.

In June the RAF AOC, Air Commodore R.M. Hill* left the country on further appointment, and was succeeded by Air Commodore Arthur Harris as from 18 July. 'Bert' Harris, the later famed leader of RAF Bomber Command, 1942-45, was under no delusion as to the frustrations and lack of any possible final solution to such a situation as existed in Palestine in 1938, having already accumulated firsthand experience of similar

* Later Air Chief Marshal Sir Roderic, KCB,MC,AFC.

circumstances in India and Iraq in previous years. His views were best summed up in his autobiography*:

> My advice to all young commanders in all Services is, whenever you see any prospect of being called out 'in aid of the civil power' in any part of the world, to get the hell out of there as quickly and as far as you can. If you fail by being too soft you will be sacked; if you succeed by being tough enough, you will certainly be told you were too tough, and you may be for it. Therefore I say that the best thing to do is to take long leave, or to get transferred, or to retire and buy yourself a farm; do anything, in fact, sooner than get involved 'in aid to the civil power'. There are only two things you can get from aiding the civil power ... brickbats and blame.

In the month that Harris arrived the general situation showed no sign of diminishing, indeed it showed an overt increase in rebel and terrorist activities. Existing plans for large reinforcement of the land forces now included strengthening the RAF in Palestine, and the first air reinforcement was the arrival of two Flights and the unit HQ of No 211 Squadron from Helwan, Egypt, becoming based initially at Ramleh. 211 Squadron, flying Hawker Hinds, had only arrived in Egypt from the UK in May 1938.

In August further reinforcement came in the shape of two Gloster Gladiator fighters from No 80 Squadron, based at Ismailia, Egypt, which arrived on 23 August and were joined by two more Gladiators in the first week of September. Two of these 80 Squadron pilots were Pilot Officer P.G. Wykeham-Barnes and Pilot Officer M.T. St J. Pattle, both of whom first fired guns in anger during several strafing attacks on rebel bands in early September before returning to Egypt later that month. Both pilots were later to achieve fame: Pattle became Squadron Leader, DFC and Bar and the RAF's highest-scoring fighter ace of the 1939-45 war and was killed in action over Greece on 20 April 1941, while Wykeham-Barnes was to eventu-

* *Bomber Offensive* by Sir A. Harris; Collins, 1947

ally retire from the RAF as Air Marshal Sir Peter Wykeham, KCB,DSO,OBE,DFC,AFC.

The 80 Squadron detachment was replaced by a Flight of No 33 Squadron (Gladiators, based at Ismailia) on 26 September, and this Flight was joined by the rest of 33 Squadron shortly after, commanded by Squadron Leader H.D. McGregor, DSO. By then a Flight of 6 Squadron, which had been based at Semakh, had been moved to the Haifa civil airport for air operations over the northern areas; the chief centre of disorders then. In September, among many minor actions, the most important event was probably that which occurred on the 15th, when a rebel force estimated 500-strong was attacked at Deir Ghassana by aircraft from Nos 6, 80 and 211 Squadrons throughout daylight hours and caused a rebel death toll of over 130, apart from unknown wounded or injured. Ground fire was heavy – several aircraft sustaining multiple bullet scars – but failed to ward off the aircraft.

By the end of September 1938 No 211 Squadron had returned to Egypt despite the continuing seriousness of the Palestine operations, leaving the onus of operations to No. 6 Squadron, recently re-equipped with Hawker Hardys and now commanded by Squadron Leader N.C. Singer, DSO (later, Air Commodore, CBE,DSO,DFC), which was still based at Ramleh but had two Flights detached to landing grounds further north. Sharing the onus from September was No 33 Squadron, while No 14 Squadron, based at Amman in Transjordan, provided important support in various ways.

Active air operations against rebel terrorists were by now virtually a daily occurrence, spread across many different locations within Palestine. On 18 October, accordingly, virtually all military and civil security forces were placed under the control of the army GOC, Lieutenant-General R.H. Haining, CB,DSO, and next day British forces occupied Jerusalem, driving out rebels formerly in residence in the city.

Operations during the autumn of 1938 brought announcements of several more gallantry awards to RAF personnel, including a DFC to Flight Officer F.C. Hopcroft (later Group Captain, DSO,DFC), a DFM to Sergeant J. Onions (later Squadron Leader, DFM and Bar), and a rare Military

Medal to AC1 J.L. Crerar, who had displayed great courage while attached to the 2nd Battalion, The Black Watch as a ground wireless operator on 13 September, when the section was ambushed eighteen miles south of Jerusalem. Another rare award was a Bar to a DFM for Sergeant J. Coggins for courage in action on 11 March 1939, his original DFM having been awarded for similar gallantry on operations in November 1938.

Apart from direct actions against any armed rebel forces, the RAF co-operated closely with the land forces in a new tactic known as 'Air Cordon', whereby complete villages, towns, even cities – Jerusalem on 18 October being an example – were 'cordoned off' by aircraft flying around the outskirts to prevent any inhabitants leaving, while the army proceeded to occupy and thoroughly search the place for rebels and any illegal arms. It proved to be a very successful tactic in late 1938 and early 1939, being used on more than 100 occasions in November-December 1938 alone. By the end of 1938 the air effort by Nos 6 and 33 Squadrons had been as follows:

	No 6 Sqn, Jan-Dec inc.	*No 33 Sqn, Sept-Dec inc.*
Flying Hrs Bombs dropped	5280 768 x 20lb 29 x 112lb	1850 –
Rounds SAA fired	62,540	92,100
Estimated rebel casualties	597	125
RAF casualties (operations)	2 pilots killed 1 pilot missing 1 pilot injured 2 AGs killed 1 AG injured	1 pilot killed 1 pilot died (injuries) 2 pilots injured
No of aircraft hit by enemy fire	41 (inc six destroyed on scene of action)	17 (inc three destroyed on scene of action)

The only other unit actively involved in late 1938, No 14 Squadron, though not directly involved in Palestine actions, mounted a prodigious number of both day and night sorties in its Wellesleys. Chiefly concerned with preventing incursions of armed Arab forces across the Transjordan and Syrian borders into Palestine, the squadron also patrolled off the Palestinian coast to prevent illegal immigrations and arms smuggling, and on several occasions undertook night bombing operations against known rebel concentrations, dropping flares and 20lb bombs, which tactic induced near-terror among some rebels unaware that aircraft could operate by night. Mention must be made too of the indefatigible crews of No 2 Armoured Car Company RAF which, by various detached sections, operated actively throughout the year, supporting both land and air actions.

By the start of 1939 rebel activities became less direct in the context of actions against security forces, resorting much more to pure terrorist tactics, although the government forces were to continue having to attack any rebel concentrations discovered. The 1939-45 war merely interrupted such activities, but as soon as the war was over Arab and Jewish terrorism – not only against each other but particularly against British occupation Services and administration – escalated rapidly as the imminent Partition Plan loomed.

Ultimately, British forces were withdrawn from their Palestinian bases, and legal mandate control by the UK government ceased on 14 May 1948 – the date on which the newly-created state of Israel came into being. Within 24 hours Israel was simultaneously invaded by forces from Lebanon, Syria, Iraq, Jordan and Egypt, supported by irregular forces, and military contingents despatched from Saudi Arabia, Morocco and the Sudan. The subsequent Arab-Jewish conflicts are now history ...

CHAPTER SEVEN

In Indian Skies

In December 1910, a mere seven years after the Wright brothers had first accomplished flight in a man-controlled, powered aeroplane, the first-ever aircraft in Indian skies arrived at Calcutta when a small trade mission from the UK-based British and Colonial Aeroplane Company (later retitled Bristol Aeroplane Company), comprised of Farnall Thurstan and a pilot, Henri Jullerot, two mechanics, and two Bristol Boxkite aeroplanes (Nos 9 and 12), disembarked. The purpose of this mission was purely commercial, intended hopefully to persuade local Indian potentates and government officials to purchase further aeroplanes from the Bristol-based firm.

Though merely civilians in 'trade', the mission was fortunate to be immediately taken under the wing of the army's local Deputy Assistant Adjutant and Quartermaster-General, a 33-year-old captain named William Sefton Brancker*, whose personal enthusiasm for aeronautics led him to arrange for the initial demonstrations of the Boxkites from the racecourse of the Calcutta Turf Club. Thus, on 6 January 1911, Henri Jullerot took off in Boxkite No 12 in front of a large vice-regal party audience – the first powered aeroplane flight ever in Indian skies outside Hindu mythology.

Nine days later that historic flight demonstration Jullerot, with Brackner as his observer, was officially requested to undertake an aerial reconnaissance over some cavalry manoeuvres in the Deccan, and successfully accomplished their task, flying some 27 miles at a height of 1,000 feet and accurately locating and reporting dispositions of the 'enemy'

* Later, Major-General Sir Sefton, KCB,AFC.

troops within two hours of take-off; a feat witnessed by among others Sir Douglas Haig, CGS, Indian Army. This practical application of aircraft in a military context particularly impressed the C-in-C, Indian Army, General Sir O'Moore Creagh who resolved to create a military flying organisation in India. By December 1913, despite official constraint on allocation of funds, his efforts had resulted in the establishment of an 'Indian Flying Corps' (*sic*) flying school at Sitapur in Uttar Pradesh. Commandant of this school was Captain S.D. Massey from the 29th Punjab Regiment and unit equipment comprised just four aircraft. Massey's orders were to offer flying instruction to selected officers (British only) of the Indian Army ' ... in Indian conditions' (*sic*), with the initial training course planned to commence in September 1914. However, the outbreak of war with Germany in August 1914 immediately nullified this intention and the first four officers scheduled to attend for instruction were sent instead to England to train under the aegis of the UK-based Royal Flying Corps, and were then attached to various RFC units in Europe.

By April 1915 the Viceroy of India, convinced of the need for military aircraft based in India, requested provision of an RFC unit from England to assist army operations in northern India – a highly sensitive area at the best of times but now exacerbated by Turkey's decision to ally itself to Germany and the Central Powers in the European struggle. To meet that request 'A' Flight of a newly-forming unit, No 31 Squadron RFC, was officially formed at South Farnborough on 11 October 1915, consisting of five officers, 44 non-commissioned ranks, and five BE aircraft all drawn from No 1 Reserve Aeroplane Squadron. This Flight was promptly despatched to India and arrived at Bombay on 26 December 1915, then moved to its first base at Nowshera, some 30 miles east of Peshawar on the north-west frontier; the first RFC unit to become resident in the sub-continent of India.

By July 1916 the unit was joined by 'B' and 'C' Flights from England, by which time 31 Squadron, now complete, was based at Muree under the command of Major C.R.S. Bradley. From early in 1916 squadron pilots had flown a number of local

BE2E of No.31 Squadron (Lt Coslet) at Hutti, NWFP, 1919.

Bristol F2B, F4611 of No.48 Squadron at Quetta, 1919., The trisected circle marking on the fuselage was a unit marking then.

sorties co-operating with the army, but true operations did not commence until October 1916 when the squadron was called on to help counter attempts by the Kuki Khel Mohmands to threaten occupation of the vital Khyber Pass. These operations set the pattern for 31 Squadron's future employment throughout World War One, consisting of such varied roles as reconnaissance, ranging for artillery guns, and occasional bombing and machine gun strafing sorties against recalcitrant tribesmen challenging the British Raj (rule).

By the summer of 1917 the squadron had grown greatly in numbers of aircraft and personnel – achieving totals of 34 Henry Farman, one Maurice Farman, one FE2b, and 26 BE aircraft – resulting in a decision to use the unit's 'D' Flight, based at Lahore, to be the nucleus of a new squadron, No 114, in September 1917. Within six months both squadrons had begun joint operations over the frontier regions which culminated (in the main) in April 1918, after which the temporarily subdued tribes maintained a fragile 'peace' for the remaining months of that year. In November 1918 the war with the Central Powers ceased with an armistice agreement, and in that month two pilots of 31 Squadron, Captain H.E. Fletcher, and Lieutenant E. Robinson, were officially gazetted as recipients of Distinguished Flying Cross (DFC) awards – the first such decorations for operations in India – although by then Robinson was dead, having been killed in a flying accident on 6 December 1917. Despite the official 'outbreak of peace' (*sic*) Nos 31 and 114 Squadrons did not suffer the fate of so many wartime-created squadrons, disbandment, but were retained in India for continuing support of the garrison armies; the description 'peace' was rarely applicable to the north-west frontier regions of India at *any* period of the nineteenth and early twentieth centuries.

One facet of the two squadrons' work throughout 1916-18 had been to locate, photograph, and report on the suitability of a variety of potential landing grounds along the frontier region; resulting in the adoption of several 'airfields' for forward operations which would soon become permanent RAF stations in some degree, albeit in many cases simply adjacent to established army forts or posts. By 1919 a number of these landing grounds possessed fundamental necessities – fuel, ammunition, *et al* – as stock for emergency operations, guarded by local army formations.

Such 'airfields' were soon to be used for further operations in early 1919 when unrest again erupted with a series of riots at Amritsar. 31 Squadron, based then at Risalpur, despatched 'C' Flight, still flying antiquated BE aircraft, to support the local army and flew several bombing and strafing sorties. These riots soon subsided but a far greater menace to peace came in May 1919 when Afghanistan declared war on India – the so-termed Third Afghan War.

This declaration had resulted from events within Afghanistan* when, in February 1919, the hereditary ruler Amir Habibulla had been assassinated and was succeeded by his son Amir Ammanulla. The latter was immediately beset on all sides by pretenders to his throne and burgeoning unrest throughout his kingdom. In order to rally his people – and divert their minds from the increasingly disturbing inner troubles – Ammanulla decided on war with India, with hostilities commencing on 6 May; a two-fold objective in that the riches of the Indian border plains had always been a coveted prize to many generations of marauding Afghans.

The Third Afghan War was destined to last just four weeks during which many thousands of Afghan Army regular troops aided by thousands more sympathetic (or opportunist) armed tribesmen were fiercely engaged at various points by British and Indian army columns supported by the few available aircraft of 31 and 114 Squadrons. Still flying mostly outdated aircraft, the RAF had nevertheless recently received a handful of more modern types, including two truly 'heavy' bombers, Handley Page 0/400, C9700 and Handley Page V/1500, J1936, this latter being named HMA *Old Carthusian*. Both bombers had arrived in India in December 1918 and January 1919 respectively, both aircraft having been *flown* all the way from England via France and Egypt.

* Afghanistan was given its independence by the 1919 Peace Conference.

Handley Page V/1500, J1936, '*HMA OLD CARTHUSIAN*', at Karachi, May 1919. It originally arrived at Lahore on 29 March 1919.

De Havilland 10, E5453, 'H' of 60 Squadron.

Pilot of the *Old Carthusian* for its epic UK-India flight had been an ex-RNAS veteran Robert ('Jock') Halley, holder of a DFC and Bar for his wartime operations and now awarded an AFC for this flight. With the eruption of the Third Afghan War Halley was ordered to fly 0/400, C9700 to Risalpur on 13 May in order to carry out bombing raids on targets in Afghanistan. At Risalpur, however the huge biplane became a victim of a gale-force storm and was wrecked at its dispersal. Halley was then ordered to fetch the V/1500, *Old Carthusian* from Lahore for such an operation, and by 23 May the Handley Page was ready for its first sortie, having had some external bomb racks (from 31 Squadron BEs) fitted to the lower wing spars for carriage of 112lb bombs, with a further sixteen 20lb bombs stowed in the dorsal cockpit for hand-dropping in conjunction with the heavier missiles. At 3 am on 24 May 'Jock' Halley took off with Lieutenant F.E.E. Villiers* as his volunteer bomb-aimer.

Halley's own account of this sortie read:

> The route lay towards the Khyber Pass and as the clearing height was about 3000 feet this meant flying around for about an hour to gain height before going over the ridge of hills. The Khyber was only dimly visible, as were a few lights at Jamrud Fort. We flew over the Kabul river and as daybreak came we reached Jallalabad, but to my horror I saw water leaking from the base of the second cylinder. I called Flight Sergeant Smith up to me and, with engines throttled back, we hurriedly conferred on the course of action. The leak came from a defective rubber connection running along the base of the six cylinders of the starboard engine but it was not possible to judge the extent of the water loss. Kabul was still 90 miles away and we had to get back. Finally with Villiers and Smith we decided to carry on, me keeping a very watchful eye on the leak as we passed through precipitous mountains.
>
> Then we came to the Jagdalak Pass which I could see high

* Later Sir Edward Villiers.

> above me. Chancing all I hauled back on the stick and just cleared the top, then dropped down the other side. Below, Villiers could see dozens of camels of a long caravan scattering in all directions, several falling over the edge of the mountain side. On the other side I saw our target ahead spread out on a vast fertile plateau. With a population of 300,000 I was not surprised at the extent of area covered by the city.
>
> Still nursing our leaking engine, we made the bomb run and our missiles achieved good results, and if that didn't frighten the inhabitants – many of whom could never have seen an aeroplane before – the sound of the *Old Carthusian* just a few hundred feet up with four engines roaring certainly did! On the return flight we again scraped over the Jagdalak Pass, helped by a lightened aircraft and a following wind. Finally the leaking engine's temperature rose so I had to close it down, but we landed safely at Risalpur after six hours in the air, being all greatly relieved.*

Four of Halley's heavier bombs had found their mark on Amir Ammanulla's palace, including demolishing a wall of the Amir's harem, and the raid so impressed the Afghans that the Amir hastily sought an armistice on 3 June, followed by a peace treaty signed on 8 August. Halley received a second Bar to his DFC for this sortie, and eventually died as Group Captain, RAF on 13 December 1979. His aircraft, the *Old Carthusian*, never flew again due to the discovery of extensive damage by termites to its wing spars, although its capacious fuselage continued to serve as a squadron office at Risalpur for several years. Though initially designed for bombing raids on Berlin from UK bases in 1918, this Kabul sortie was in fact the only war operation ever flown by any Handley Page V/1500.

Even as the Third Afghan War was in progress the first peacetime establishments for the infant RAF were beginning to be planned and implemented, with particular emphasis upon the various overseas territories under British imperial and

* Quoted in *First in the Indian Skies* by N. Franks; RAF Collection, 1981.

Bristol F2B of No.20 Squadron.

Fort Dardoni, viewed on 4 October 1924. In background is the original site of Fort Miranshah, which was enlarged by transferring all facilities of Dardoni.

mandated control. As part of that commitment three squadrons based in Europe were now allotted to India in addition to the two resident units already there (Nos 31 and 114 Squadrons). These extra units were Nos 20, 48 and 99 Squadrons and all three units had reached India by the close of June 1919, being based initially at Risalpur (20), Quetta (48) and Ambala (99), and equipped with Bristol F2Bs (20 and 48) and De Havilland 9As (99). Shortly after a fourth 'new' unit arrived in India when No 97 Squadron (De Havilland 10 twin-engined bombers) docked at Bombay on 12 August and became based at Allahabad and Lahore (HQ) initially.

This tripling of RAF strength in India might have suggested an upgrading in importance of the air arm, but in practice the whole RAF there remained merely of Group status whose overall RAF commander in mid-1919 was Wing Commander R.P. Mills, MC,AFC. Nevertheless, although still only a group under the aegis of the C-in-C, Army in India, its status was partially boosted on 20 September 1919 when the RAF commander's appointment was raised to Air Commodore, the first such air officer being Air Commodore (later Air Vice Marshal Sir) T.I. Webb-Bowen, CB,CMG.. Within a few weeks of Webb-Bowen taking up his appointment further changes in the India units took place. For reasons and logic known only to the bureaucratic minds closeted in the Whitehall corridors of paper power 'improved'* the RAF in India by calmly retitling several units. On 1 February 1920 Nos 48 and 114 Squadrons became Nos 5 and 28 respectively, while two months later on 1 April Nos 97 and 99 Squadrons were renumbered as Nos 60 and 27 Squadrons respectively. In addition two 'new' units were recreated when, on 20 January 1920 No 1 Squadron was officially reformed at Risalpur, and on 1 February No 3 Squadron was 'reborn' at Ambala; both squadrons being equipped with Sopwith Snipe single-seat fighters. Thus on 1 April 1920 RAF squadrons in India were:

* Presumably according with an old RAF adage 'If you change the label, it becomes more efficient' ...

Squadron	*Main Base*	*Officer Commanding*
1	Risalpur	Sqn Ldr J B Graham, MC, AFC
3	Ambala	Sqn Ldr G G A Williams
5	Quetta	Sqn Ldr P C Maltby, DSO,AFC
20	Risalpur	Sqn Ldr J H S Tyssen, MC
27	Mianwali	Sqn Ldr R J Mounsey, OBE
28	Ambala	Sqn Ldr R J F Barton, OBE
31	Risalpur	Sqn Ldr A L Neale, MC
60	Risalpur	Sqn Ldr J C Quinnell, DFC

In support of the firstline squadrons were various depots, aircraft parks and headquarters units, apart from various small detachments scattered around the sub-continent.

All these changes took place in a period of widespread unrest along the northern frontier of India when all squadrons were being called upon for daily support of the army formations attempting to restore law and order to the region. This latest example of tribal rebellion was officially deemed to have ceased on 7 May 1920, and the work of the RAF was singled out by the official historian for special praise, with the added a significant recommendation 'officers of the land forces would find their labour well repaid if they undertook the study of this important auxiliary to the art of tactics.'* In the event this recommendation fell on deaf ears among the army and political hierarchy in India who continued to regard the century-old employment of large, cumbrous and expensive army columns as the prime solution to all tribal insurrections.

The recreation of the two Snipe squadrons, Nos 1 and 3, had no appreciable effect on the prime operational tasks of the RAF in India; indeed, due to lack of spares – the stores officer of No 1 Squadron at one stage had to obtain necessary nuts, bolts and split pins from the local bazaar – and no given operational role, both squadrons' Snipes spent the majority of days grounded. The squadrons' officers' morale, already low due to the lack of flying, was not enhanced by an order from the AOC,

* *Operations in Waziristan, 1919-20*, 2nd Edn; HMSO, 1924.

Webb-Bowen for all officers' pay and allowances in India to be reduced. The reason given was that the AOC considered too many officers were running up too high bar bills in their messes; a decision hardly mollified by a simultaneous decision to raise substantially the AOC's own salary ... In the event both Snipe units were not to remain long in India. In late April 1921 No 1 Squadron received orders to move en bloc to a newly-constructed aerodrome at Hinaidi, Iraq, while No 3 Squadron was simply disbanded at Ambala on 1 October 1921, being reformed officially on the same date at Leuchars, Scotland for a coastal recce role initially.

For all operational intents and purposes the RAF strength in India was to remain six squadrons until late 1928, despite the various official recommendations and reports submitted by visiting and resident senior RAF officers in the interim years. The overall task for those squadrons can best be appreciated by both the background to the seldom-ceasing tribal wars and revolts along the Indian north-west frontier, and at an esoteric level the difficult status of the RAF vis-à-vis the army, and the hangar-floor myriad practical problems of attempting to fulfil the various operational roles expected of the six squadrons. The north-west border with Afghanistan had been the centre of insurrections and prolonged bloody conflicts for nearly a century before the RAF/RFC had appeared in the area.

Throughout most of the nineteenth century the main source of trouble for the British in India had been Afghanistan which, with no small encouragement from Russian ambitions to expand the Czar's empire eastwards, instigated numerous wars and rebellions spreading well into northern India. In 1893 an agreement was finally concluded with Amir Abdur Rahman of Kabul to establish a recognised frontier boundary between Afghanistan and 'British India'; this political boundary becoming known as the Durand Line after Sir Mortimer Durand, head of the boundary commission.

At that same time a second 'administrative' border line was confirmed, some 50 miles east and roughly parallel to the Durand Line and extending for about 450 miles north to south. It was agreed that all inhabitants of the territories *between* these two border lines were free from administrative control, i.e. able to live their lives as they saw fit without outside interference. This 'free' zone, some 27,000 square miles of wild mountainous country virtually inaccessible to modern transport, held a population of at least two million tribesmen – Mohmands, Afridis, Wazirs, Mahsuds, and other lesser tribes, collectively and generically referred to as Pathans. Surrounding them, both in Afghanistan and within British-administered frontier territories were a further eight or nine million tribesmen.

These Pathans, of whatever individual tribe or origin, held several characteristics in common. A warrior race, proud, fierce and cruel to whom possession of a rifle since boyhood was the mark of a true man, the Pathans were deeply religious in their faith, Islam, with its promise of an honoured place in Paradise for those who died in battle. All lived by a strict code of honour – *Pakhtunwali* – which decreed among other obligations safety and protection for any fugitives, hospitality at all times even to a hated foe, and particularly the avenging of any insult (real or even imagined) to individuals, families or tribes. Theft by any 'outsider' of *zar, zan,* or *zamin* (gold, women, or land) immediately entailed a feud in blood. Above all, Pathans acknowledged no man as their master in any matter except religion.

In the latter, however, their belief in the teachings of their *mullahs* (religious teachers/leaders) and occasional fanatical *fakirs* (holy men) was total, especially if such men advocated a *Jehad* (holy war) against the infidel. Thus, born and raised to regard war and fighting as their inheritance, sustained by absolute adherence to their religious beliefs, and possessing an unrivalled knowledge of every cranny of their mountain habitat, the Pathans were the finest guerilla fighters in the world.

This regard for the honourable, masculine 'profession' of war was even extended to many of their non-Moslem enemies, particularly the British, and occasionally resulted in (to the British) somewhat puzzling and even humorous episodes. For merely one example, when a particular campaign had been concluded in the 1920s, Pathans applied in all seriousness for

Bristol F2Bs on detachment at Dardoni. Note the fort's name spelt out in white stones at centre left.

their awards of the British India General Service Medal with its appropriate campaign clasp. When denied this the Pathans failed to understand the logic – had they not performed a great service fighting bravely and thereby providing valuable service in training for the King-Emperor's army?

On 9 November 1901 a recently arrived Viceroy of India, Lord George Curzon, brought into existence the new North-West Frontier Province directly responsible to the government of India which brought all the frontier areas under one head, a Chief Commissioner, for political and administrative control. In the following year, at Curzon's request, a new C-in-C India arrived, Lord Kitchener of Khartoum, who immediately reorganised the army with the prime principle that the army's chief role was to be defence of the North-West Frontier against external aggression, leaving internal security in India itself to the established police forces. The chief menace to peace was still Russian presence in Kabul and other bordering countries like Tibet and Persia. By 1907, however, Russia's defeat in its war with Japan led to the 1907 Convention of Anglo-Russian talks which finally acknowledged Afghanistan as being solely a British concern, although Britain also agreed not to interfere in Afghan *internal* affairs.

By the outbreak of war between Britain and Germany in 1914 Afghanistan was ruled by Amir Habibulla who, despite Turco-German overtures and some short-lived tribal eruptions (provoked by Turkish agents via certain fanatical *mullahs*), remained neutral and pro-British, thereby ensuring peace along India's North-West Frontier Province until the close of the European conflict. Once the war was over Habibulla approached the British government seeking some reward for his help in maintaining the peace, in particular seeking independence for Afghanistan in its foreign affairs. Before any such negotiations could commence Habibulla was assassinated, his brother Nasrulla assumed his throne, only to be arrested and life-imprisoned by Habibulla's third son, Ammanulla. In May 1919 Amanulla moved his army up to the Durand Line, made incursions over this border, and thereby precipitated the Third Afghan War which, as described earlier, culminated in a peace treaty being signed on 8 August 1919 in Rawalpindi.

For RAF personnel stationed in India during the 1920s life was peculiarly antiquated in many respects. Steeped in centuries of long-held traditional customs, the rituals of social life under the British Raj contrasted quaintly with more modern requirements of progress. To the newly-arrived on the Indian scene such customs gave a distinct impression of stepping back in time to the Victorian era, with all its emphasis on the trivia of gentlemanly conduct and graceful living. To the professional airman such a mode of living was linked directly in vintage to the equipment he was expected to employ. His aircraft – Bristol F2Bs, DH9As, and DH 10s – had been designed for wartime exigencies for use in European climates, and were now expected to maintain regular operational tasks over terrain and in climatic conditions never envisaged by their designers.

Moreover, the back-up maintenance for the aircraft suffered from a near-complete lack of spares replacements and it soon became a common practice to fit locally-manufactured wooden 'wheels' to some aircraft in order to reserve rubber tyres for those machines actually undertaking operations. Such shortages also led to cannibalisation of two or even three 'non-operational' aircraft in order to bring a single aeroplane up to flying fitness.

The fundamental reason for such a low level of operational capability was simply finance – or rather lack of it. Whereas the remainder of the RAF was regarded as a separate Service with its own annual budget allotted by the Treasury, the RAF in India was still merely one of many expenditure items in the Army budget which was controlled by the C-in-C, India, an army member of the Viceroy's Council. The most senior RAF officer was junior in rank to his army commander and thus had little weight in decisions on policy or financing. To quote MRAF Sir John Slessor, who served as a Flight commander with No 20 Squadron at that period:

> The RAF in India suffered less from the malevolence of a tribal enemy than the depredations of the Government of India ... I think it was inevitable that among senior advisers to the Viceroy, the combinations of ignorance about Air

Bristol F2B, J6782 of 31 Squadron (pilot Flg Off Ford) getting airborne from Dardoni 1923.

CLOSE STUFF. Bristol F2B, F4839 of 31 Squadron (Flg O Fresson & LAC France) in ver close formation over Delhi.

DH9A of No.27 Squadron (Fl Offs Holdway and F G S Mitchell) about to leav Miranshah for a bombing sortie, 1924.

> matters, ingrained with tradition and the Englishmen's national suspicion of anything new should have had the result that, when cuts in military expenditure were required, they should fall upon this new Service, which no one understood.*

The offical cessation of the Third Afghan War appeared to bring a modicum of peace to the northern-most area of the NWFP, albeit not stemming the almost routine trans-border raiding by small Afghan tribes which were a constant menace to law and order. Further south in Waziristan and Mahsud countries, however, further widespread tribal rebellion, was followed in late 1919 and early 1920 by yet more Afghan invasion of British-administered Indian zones on the frontier.

All RAF aircraft readily available were committed to repelling these uprisings, one example being the so-termed Battle of Mandanna Hill, 20-25 December 1919 when four F2Bs, six DH9As, and three DH 10s maintained a daily bombing/strafing offensive against the invading Afghans. Such operations were by no means as one-sided as might be thought, exemplified by a raid against rebels in the Ahnai Jangi Gorge on 14 January 1920 which resulted in three Bristol F2Bs being shot down by the deadly accurate Pathan rifle fire; two F2Bs were wrecked and their crews killed, while the third crashed without serious injury to its crew.

The fate of any RAF crew forced down in tribal territory when engaged on operations was usually precarious. All air crews carried 'safety certificates', or as these were commonly nicknamed 'gooli chits', which were documents, printed in the native Urdu and Pushtu language, guaranteeing ransom for the return *un*harmed of any British personnel. However, the actual treatment of such captured fliers depended greatly on the individual circumstances. Had a crew been bombing and gunning their captors only minutes before then at least a hard beating or other rough handling could be expected. And if the tribesmen were in an unforgiving mood the prisoners might be

* *The Central Blue;* Cassell, 1956.

handed over to the tribe's females whose 'attentions' might well include crude castration without benefit of anaesthetic ... hence the nickname of the 'safety' chits! The male Pathan's capability for dire cruelty to any foe was certainly outmatched by his female counterpart; the epitome of Kipling's description of the female of the species being deadlier than the male ...

The years 1920-22 proved just as turbulent as previous years, particularly in Waziristan where, in 1923, it was decided to establish a permanent garrison of regular army troops at Razmak in the heart of that territory, with an increase in supporting native and British army brigades and battalions based along lines running north and south of Razmak. Meantime the handful of RAF squadrons attempted doggedly to provide as much air co-operation as practicable despite steadily declining maintenance facilities due to the continuing niggardly finance allotted by a parsimonious Indian government. Each squadron was nominally established for twelve aircraft for firstline operations, but the actual availability on the squadrons rarely reached such a figure. For one example, the actual disposition of RAF squadrons in India as at 30 June 1922 was:

Squadron	*No of aircraft*	*Main base*
5	9	Quetta
20	9	Ambala
27	7	Risalpur
28	7	Parachinar
31	8	Peshawar
60	5	Risalpur

It should be noted that these figures represented actual aircraft, *not* necessarily fully serviceable machines ready for operations. By that date the aircraft maintenance situation had deteriorated to a scandalous level. As the commander of No 31 Squadron, Squadron Leader Arthur Harris* remembered:

* Later MRAF Sir Arthur Harris, Bt, GCB,OBE,AFC,LLD, AOC-in-C, RAF Bomber Command, 1942-45.

> We lacked practically everything which an Air Force squadron anywhere else would regard as essential for maintaining its aircraft. We came under the Army Vote and, as a result, we got little of everything and much of what we got was useless as, for instance, the photographic materials and chemicals, which had all deteriorated until they were no good. We had single-ignition Rolls Royce Falcon engines (dependent on one contact breaker) when new dual-ignition engines were being sold for a song as surplus in England!

A year later Flight Lieutenant Strang Graham (later Air Commodore, MC,GM) joined 27 Squadron and found the situation unaltered:

> The aircraft and engines were really very old and constantly requiring attention. There were numerous cases of a connecting rod going through the sump of the engine and some very nasty and horrible forced landings resulted. Often as aircraft were taxying out from the hangars, a tyre would puncture on the sharp camel thorn on the ground with the result that we had many patches on our inner tyres.*

Such a dangerous situation applied to all squadrons – some Bristol F2Bs of 20 Squadron still had patched bullet holes from 1918 in their fabric – and among many officers who privately complained was Flying Officer R.B. Luard of 27 Squadron who spared no details in his letters to his father, a retired colonel, in England.

Luard senior mounted a campaign of protesting letters to the national press and leading aeronautical magazines, damning Lloyd George and his government for permitting such a state of affairs, and the consequent crescendo of criticism eventually reached the floor of the House of Commons provoking many very heated exchanges across the Speaker's chair.

Bowing to the storm, Lloyd George ordered an official inquiry to be sent to India, headed by Air Vice-Marshal Sir John Salmond in the early summer of 1922. On arrival, with his

* Correspondence with the author 1966.

DH9A, E8674 of No.60 Squadron, ca. 1923, which later served as 'G' of 27 Squadron.

Bristol F2Bs of 5 Squadron keeping close company over typical NWFP territory.

characteristic tact and seemingly boundless energy, Salmond promptly made a tour, talking with high-ranking army officers, political figures, and especially the RAF squadrons, then in August 1922 submitted his findings and earnest recommendations in a highly detailed 37-page report to the Viceroy of India.

By a tragic coincidence at almost the same moment as Salmond's report was received by the Viceroy, Flying Officer Luard was killed in a flying accident. On 17 August 1922, accompanied by Flying Officer D.C. Duncan, Luard took off in a DH9A in company with three Bristol F2Bs from Dardoni for a bombing raid. At 150 feet after take-off the DH9A spun into the ground, burst into flames and killed both occupants. The subsequent court of inquiry into this crash attributed the reason to 'Pilot error' ...

Salmond's report, a highly detailed and deeply researched document, was divided into two parts: Part One dealt with past examples of RAF successful employment in other countries and the possibilities of the Service's use in India; while Part Two described in plain statistics and logic the existing low state of efficiency of the RAF in India, and recommendations for the administrative measures needed to maintain the RAF as an efficient, economic fighting force. Salmond's summary of the existing RAF in India was succinct:

> It is with regret that I have to report that the Royal Air Force in India is to all intents and purposes non-existent as a fighting force at this date. The number of aircraft on the authorised establishment is 70; of these two-thirds or 46 should be constantly serviceable in any climate. In the Royal Air Force in India on 23rd August 1922, the total number shown as serviceable was 7 (or 15 per cent of expectation) and of this number a percentage are so old and decrepit that they should have been already struck off charge, while some are flying without the incorporation of technical equipment essential to safety.*

* 'Report by AVM Sir J. Salmond on the Royal Air Force in India,' August 1922; HMSO.

With a constant emphasis on the economic possibilities of a properly equipped and manned RAF in India, Salmond's report made, among other recommendations, representations for provision of two more squadrons to be added to the existing six, increases in personnel, accommodation, and financial provisions, with an upgrading of the senior RAF officer's status and rank to air vice-marshal.

One additional but crucial recommendation by Salmond was a plea for the RAF in India to be allotted a financial budget *separated* from the Army Budget as in previous years, adding, 'Consequent on this increase (*referring to the additional two squadrons he had recommended*) the utilisation of the Royal Air Force as the sole punitive weapon for the control of tribesmen in Waziristan *with resultant economy in both the numbers and cost of maintenance of troops of occupation, and in the cost of punitive operations*'* Clearly, Salmond was pursuing the Trenchard principle of air control *without* the need for large, expensive army garrison troops in occupation about to be exemplified in Iraq and in accordance with the findings of the 1921 Cairo Conference in respect of overseas RAF employment.

The reaction to Salmond's report was provisional approval by both the Indian government and indeed by the army C-in-C; in both cases with the reservation of how and from where the necessary extra finance could be provided. Yet within a year, and indeed for a decade thereafter, air power as a factor of overall operations in India was ignored by successive army and Vice-regal committees when policies were debated and proposed. Even the two extra squadrons recommended by Salmond – and agreed by the authorities in 1922 – were not actually despatched until six years later.

Evidence of that neglect of the RAF in India is succinctly summarised in part of a note to the Air Staff by the AOC, RAF India in January 1931, when he stated: 'In the holding of the Frontier and in Indian defence generally our present organisation is still substantially that of pre-aircraft days, and does not take account of the new arm and new methods, and

* Salmond's emphatic italics in his report.

that we could and should save many lives and much expenditure by increasing the strength of the air forces and making more use of them' – in essence a reiteration of Salmond's recommendations of nine years before.

The initial near-token approval of Salmond's report's overall recommendations had little, if any, practical effect on the squadrons' maintenance problems. In the context of actual aircraft, No 31 Squadron had at last exchanged its ancient BE machines for Bristol F2Bs in late 1919, while 60 Squadron, reduced to merely two unserviceable DH 10s by the end of February 1923, began re-equipping with DH9As from April and was fully equipped with Ninaks by the autumn of that year.

Thus by late 1923 the six squadrons were flying either DH9As (27 and 60) or Bristol F2Bs (5,20,28 and 31), although the majority of these were refurbished wartime aircraft lacking many more up-to-date modifications. Replacement spare parts remained virtually unobtainable, leaving each squadron's servicing airmen to continue their familiar routine of local improvisation and cannibalisation in order to produce reasonably airworthy aircraft for actual operations.

Such operations at this period, 1922-23, were virtually a daily routine unless weather conditions prevented flying, and a more detailed look at a typical month, January 1923, will give a fair picture of the operational effort made by the RAF. On 1 January the total strength of aircraft held on the six squadrons was 56 (18 DH9As and 38 F2Bs), of which some 30 per cent were unfit for operational flying:

Operational sorties flown		71
Bombs dropped:	230lb	166
	112lb	711
	120lb	880
	Baby Incendiary	2351
SAA rounds fired		19,384
Photographic plates exposed		169
Total operational hrs flown		626
Casualties		2 aircraft

The two aircraft lost were a DH9A from 27 Squadron (Flight Officer R.J.M. De St Leger/AC1 Fearn) which had an engine failure and force-landed just east of Mandesh on 21 January. Its crew were both uninjured and, despite some anxious moments in the hands of the tribesmen they had been so recently attacking, were later escorted unharmed back to Ladha. The other lost aircraft was a Bristol F2B from 28 Squadron, crewed by Flying Officer R.M. Foster, DFC/L.H. Ridley, on 24 January.

Foster had set out from Dardoni at 1630 hours that day to bomb villages south of Bakri Punga with his F2B carrying a dozen 20lb bombs on its underwing racks and a box of Baby Incendiary bombs inside the rear cockpit for hand distribution. While attacking one village the bottom of the box of incendiaries broke spilling the bombs into the fuselage with eight or nine of these immediately igniting. Ridley managed to throw five or six of the burning incendiaries overboard, sustaining severe burns in the process, but could not reach others which had rolled under the rear petrol tank.

Foster immediately took the F2B down and force-landed as quickly as possible, the aircraft turning over on landing though neither man received any injuries from this mishap. Getting well away from the crash, both men saw the F2B blaze into flames and within minutes it was wrecked in an explosion. Shortly after the two officers were captured, promptly beaten up, and had most of their clothing stolen, before being taken individually to different tribal villages. For the following few nights each man was tied up to prevent any attempt of escape, while Foster was occasionally taken and tied to a post in the centre of 'his' village in the daytime whenever the villagers were warned of any approaching aeroplane, being informed in no uncertain manner that should the village be bombed he would be shot immediately.

Within the first week the village was visited by a political department employee who persuaded the tribesmen to cease tying Foster up each night, while fresh clothing was sent for Foster and Ridley from Sorarogha (although only 20 per cent of this actually reached the pair). The second and third weeks of their captivity brought better treatment, with no enmity being shown by their captors, and they were finally escorted back to Sorarogha on 12 February from where they were air-lifted back to their squadron. Foster remained in the RAF as his career, ultimately rising to become Air Chief Marshal Sir Robert M. Foster, KCB,CBE,DFC.

Apart from all the common hazards of flying over mountainous territories, allied to the possibilities of physical torture or death at the hands of vengeful Pathans in the event of capture, RAF crews also faced dangers from the constantly varying weather conditions over the North-West Frontier Province. In July 1923, for example, during a series of raids near Razmak, dense clouds suddenly shrouded the hills surrounding the target zone causing several crashes, including two Bristol F2Bs from 20 Squadron whose four crew members were killed. With such variances in landing grounds as the 6,600 feet of Razmak and the 1000-1500 feet of the Peshawar Plain, and taking into account that a normal bombing height could be as high as 12,000 feet above sea level, an F2B or Ninak pilot needed many months of experience of local flying conditions before he could be considered truly 'operational'.

There were also the extremes encountered in temperatures which could range from a searing 120-degrees in the (theoretical) shade to below freezing point on forward landing grounds high in the mountains. Pilots new to the Indian conditions were (if time permitted) often required to undertake a series of solo flights to height to acclimatise them, without passengers until they had been deemed ready to assume the responsibility of other men's lives – especially ground crew members who normally occupied the gunner's rear cockpit. One such 'fresh' pilot, on querying rather indignantly why was told somewhat cynically, 'Pilots we can always get, laddie, but good fitters are bloody hard to find!'

CHAPTER EIGHT

'Pink's War'

Although relatively a 'peaceful' year on the North-West Frontier Province of India, 1924 was by no means completely free of air operational activities, the most important of these taking place in July when all squadrons took part in bombing and reconnaissance sorties against sections of the Mahsud tribes in South Waziristan. This campaign was brief and by October 1924 most of the rebelling tribes had agreed peace terms with their Political Agents. However, a particular tribe, the Abdur Rahman Khel, with support from the Guri Khel, Maresai and Faridais, refused to submit to governmental terms, and in January 1925 began a series of raids against army Scout posts at Manzai, Spli Toi and Gomal.

As these forays became prolonged, Air Vice-Marshal Sir Edward Ellington, who had succeeded AVM P.W. Game as AOC, India in November 1923, made a significant decision – to instigate punitive operations against the rebels by the RAF *without* the customary army columns in ground support. It was to be the first occasion in the NWFP's history when the RAF was to conduct completely independent operations and, as such, the progress and outcome were to be closely scrutinised by army and civil authorities in view of future policy.

On 20 February 1925 Wing Commander R.C.M. Pink, the officer commanding No 2 (India) Wing, was flown in a DH9A of 60 Squadron, piloted by Flight Lieutenant J.W. Baker (later, Air Chief Marshal Sir John Baker) to Rawalpindi for a conference with the Northern Command HQ commanders and the forthcoming operations were agreed in principle between Pink and the army commanders, and five days later the customary warning leaflets were dropped among the rebel tribes' camps. Meanwhile Pink began deploying his forces, and on 3 March eight aircraft each from 27 and 60 Squadrons moved to Miranshah Fort, some 120 miles from Kohat – the first two squadrons to actually occupy Miranshah although the Fort, headquarters for the Tochi Scouts, had been in use by the army for a number of years.

The decision to extend Miranshah's boundaries to include a permanent landing ground and accommodation for aircraft had resulted from the closing and demolition of Fort Dardoni, a nearby forward army post which had been utilised by the RAF on occasion, in late 1924 and all finance diverted to improving the facilities at Miranshah. Literally a fort in the Beau Geste character, Miranshah might well be summed up in the words of an anonymous poet;

> Steep hills, red rocks, grim boulders for the feet that do not falter.
> Clear crags and slender crevices, for eyes that see afar,
> A fight for life 'twixt man and man, that ages cannot alter –
> This prospect I would show you from on high by Miranshah.

On 7 March 'A' Flight of 60 Squadron joined its brother Flights at Miranshah, while Pink established his operational HQ at another fort, Tank (*pronounced 'Tonk'*) along with 10 Bristol F2Bs of 5 Squadron, strengthened by one Flight of 20 Squadron's F2Bs on 18 March. Pink wasted no time, and once Miranshah had received its squadrons he flew to the fort from Tank to brief all personnel on the imminent operations – in itself a somewhat novel procedure at that time. Seating *all* crews, air *and* ground, in a semi-circle around him Pink proceeded to explain in detail the tactics and objectives intended – to such good effect that on concluding his talk the whole audience, quite spontaneously rose to their feet and actually clapped their applause! This unprecedented gesture of appreciation momentarily took Pink aback – in the words of one NCO present, 'Pink became scarlet – but I don't think he was displeased really' ...

On 9 March proper operations commenced, with attacks on all sections concerned but with a main concentration on the

Abdur Rahman Khel tribes in revolt. The target areas varied between three and six thousand feet above sea level, necessitating aircraft with full war-loads to restrict fuel states to roughly 60 per cent in order to gain bombing heights. At Tank all targets were numbered on a master map with specific numbers allocated to the Miranshah squadrons and the remainder to 5 Squadron at Tank. Actual raids were to be carried out in Flight formations where possible, with 112lb and 230lb bombs being released from 3,000 feet in one salvo on a signal from the formation leader. In addition pairs of Ninaks, loaded with 112lb and 20lb bombs were despatched for lower-level harassing of tribe strongholds and caves.

By 13 March the tribesmen indicated that they were willing to comply with government demands and bombing ceased for a day, but was resumed next day when the rebels failed to fulfil their promises. The raids were not without loss. On 21 March Flying Oficers N.C. Hayter-Hames/E.J. Dashwood, crewing a DH9A from 27 Squadron, were the target for some very accurate rifle-fire near their target, and the Ninak spun down and crashed, then erupted in flames. Hayter-Hames was killed in the crash while Dashwood was thrown clear. Dashwood immediately went into the burning wreck attempting to extricate his pilot but suffered serious burns. Dashwood was then taken in hand by some friendly Guri Khel who lavished elaborate care on the mortally injured man, even slaughtering several of their precious goats and using the goat fat and skins to wrap the dying Dashwood – an example of a form of chivalrous mercy for any brave man sometimes displayed by the mountain tribesmen even to foes.* Dashwood died shortly after, and the bodies of both men were eventually brought to Sorarogha by 25 March and then interred with full military honours at Dera Ishmael Khan cemetery.

On 30 March a Bristol F2B of 31 Squadron, fitted out for night flying, was flown from Ambala to Tank and then carried out its first bombing raid by night on 4 April with visible success. Encouraged by this new tactic two more of 31

* Although it should be realised that the tribes' attitude to all native troops was always cruel since they regarded such men as traitors.

Wing Commander (later Air Cdre) R C M Pink, CBE alighting from DH9A, E8723 of 27 Squadron. Note 27 Squadron's Green Elephant 'badge' on Ninak's fin.

Fort Tank, NWFP India.

Miranshah bomb dump, with 25lb Cooper (nearest) and 112lb RL bombs (rt).

Fusing 112lb RL bombs, Miranshah.

BOMBS AWAY. Single 230lb RL bomb on its way to target from a 31 Squadron aircraft.

Squadron's F2Bs joined their companion at Tank for further night sorties, these three pilots being Flying Officers Payne, Combe, and Donald Hardman (later, Air Chief Marshal Sir Donald, GBE,KCB,DFC). The arrival of the latter pair of F2Bs necessitated a partial reorganisation of the operational force which was now redeployed:

Operational HQ	Tank
1 Flt 31 Sqn (F2B)	Tank
1 Flt 5 Sqn (F2B)	Miranshah
1 Flt 20 Sqn (F2B)	Miranshah
2 Flts 27 Sqn (DH9A)	Miranshah
2 Flts 60 Sqn (DH9A(	Miranshah
Searchlight parties	Khirgi and Sorarogha

On 4 April Squadron Leader T.F. Hazell, who had only recently been appointed OC 60 Squadron with effect from 30 March, took part in his first raid but shortly after taking off from Miranshah he noticed the engine cowling of his DH9A coming loose. Jettisoning his two 230lb bombs – which landed near an army scout post to the alarm of its troops – Hazell decided to land as quickly as possible and chose Sorarogha where its sloping landing strip ended abruptly in a sheer drop into a deep *nullah* (valley). With no option but to land *down* the sloping strip Hazell skilfully ran his Ninak into a stone breastwork on the very edge of the precipice. The DH9A was a write-off but Hazell and his petrified gunner walked away from the wreck with mere bruises.

Five days later Flight Lieutenant R.C. Savery of 60 Squadron repeated this hazardous form of forced landing at Sorarogha with similar results; while on 15 April a third 60 Squadron DH9A force-landed with engine trouble in open country.

On 9 April an afternoon patrol spotted a large collection of Faridai tribesmen on the move in open country and at once attacked, at the same time calling for additional aircraft from Miranshah to back up the harassment. Considerable casualties were caused during the initial bombing and strafing but then the weather broke completely nullifying completion of this opportunity – the only occasion when a large concentration of rebels had been caught in open country. By now many 'friendly' tribes were beginning to refuse sanctuary to the rebels, having been warned that their villages and flocks would be bombed if they gave shelter to 'unfriendly' tribes.

In the latter weeks of April several rebel tribes began overtures for a cessation of the hostilities, though few of these achieved any solution due to internal disagreements among the various rebel forces. Finally on 1 May the rebel leaders agreed to government terms at a *jirgah* (meeting or conference) held at Jandola.

Accordingly RAF operations ceased, having been undertaken by day and by night over a period of 54 days almost continuously and accumulating an overall total of over 2700 flying hours. An official dispatch on these operations shortly after ended with the comment:

> This is the first occasion in India that the RAF has been used independently of the Army for dealing with a situation which has got beyond the resources of the political officers. It is at present too early to judge how lasting will be the effect or how permanent will be the impression of this display of air power on the stubborn tribesmen of the North-West Frontier, but it is claimed that in the RAF the Government of India have a weapon which is much more economical in men and money and more merciful in its action than other forms of armed force for dealing with problems which arise beyond the administrative border.

Several gallantry awards were made to individuals, including a DFC to Flight Lieutenant John Baker and a DFM to Sergeant Hawkins, both of 60 Squadron and a DSO to Squadron Leader A.J. Capel/OC 5 Squadron (later Air Commodore, DSO,DFC), but all personnel who had served under the command of Wing Commander Pink during the period of 9 March-1 May 1925 inclusive became entitled to the India General Service Medal, 1908 with a Clasp 'Waziristan, 1925'. This award, announced by the Air Ministry in December 1926, was only entitled to be

Bristol F2B, E2421, 'C' of 5 Squadron returning to Tank after raiding Jelal Khel, 1925 during 'Pink's War'.

Line-up of DH9As of 27 Squadron.

claimed by totals of 46 officers and 214 non-commissioned airmen – the rarest award for any Indian operational campaign apart from pure gallantry decorations.

Wing Commander Richard Charles Montagu Pink, already holder of a CBE, was granted accelerated promotion to Group Captain as a reward for his masterly handling of the campaign, apart from being accorded a form of immortality in RAF annals by having these operations thereafter referred to as 'Pink's War'. His personal reactions to that period were part-summarised in a piece of cynical rhyme he had composed during his many flights over the tribal areas. Titled 'Waziristan 1925', it ran:

Oh big 9A, oh big 9A
What are you doing down Razmak way?
Why these bombs? This pomp of war?
Surely your home's at Risalpur?

Chorus
Pomp be damned, don't make me laugh
There ain't no pomp on this damned strafe
I'm bombing hell from the local Khan
To prove there's PEACE in Waziristan

Oh smart Brisfit, Oh smart Brisfit*
What are you doing in fighting kit?
I'm praying hard I'll avoid a konk
On offensive patrols from a sink called Tank.

Chorus
Up the gorges and down Spli Toi
Sniped to blazes but attaboy
They call it war on the banks of the Marne
But bless you its PEACE in Waziristan.

If they ask me – what shall I say
To the folks back home England way?

* Post-1918 nickname for the Bristol F2B

Chorus

Don't you worry there's nought to tell
'Cept work and fly and bomb like hell
With hills above us and hills below
And rocks to fill where the hills won't go
Nice soft sitting for those who crash
But WAR you call it? – don't talk trash
War's a rumour, war's a yarn
This is the PEACE of Waziristan.

Returning to England in early 1926, Pink held an administrative staff post in the newly-formed Air Defence of Great Britain Command, then in May 1929 was appointed to command RAF Station, Manston. In July 1931 he was promoted to Air Commodore but was taken ill shortly after and eventually died of cancer at the Princess Mary's RAF hospital, Halton on 7 March 1932. Highly respected, genuinely popular with all ranks who served with him, flexible in thought and courageous in decision, Richard Pink's loss to the RAF was truly tragic,

'GOLDEN TRINITY'. Pilot, Rigger & Fitter; in this case, Flt Lt 'Chas' Allen (later, AVM), with Cpl Croft (L) and unidentified Erk from 27 Squadron.

depriving the Service of a potentially brilliant future Chief of Air Staff in many senior RAF officers' opinions.

Despite a certain amount of nit-picking in subsequent official Army reports – mostly displaying a barely-hidden anxiety that the Army's long-standing 'preserves' might now be poached by this upstart air arm – the legacy of 'Pink's War' could fairly be said to be a slow, grudging change in the civil hierarchy's attitudes to the RAF in India. In the 54-days' operations the RAF had yet again, demonstrated the efficacy of employing pure air power/control in place of the traditional expensive (in money and men) army methods of countering tribal revolution and dissent along the North-West Frontier Province.

CHAPTER NINE

Over the 'Grim'*

With 'Pink's War' brought to a satisfactory conclusion the squadrons were able to return to a more settled routine of more peaceful duties, chief among these being aerial photo-reconnaissance – a permanent adjunct to their operational roles at any period – and in particular a considerable amount of vertical photography attempting to map with accuracy the wild wastelands of the North-West Frontier. Other normal routine chores included regular contact and delivery flights, usually by individual aircraft, to outlying army Scout posts scattered among the mountains and valleys. A diminution of tribal actions on any serious scale was to extend to early 1927, although few weeks passed without some minor eruption due, usually, to inter-tribal disputes over land or possessions rather than any specific anti-British movement.

Taking advantage of this relatively 'peaceful' era the RAF proceeded to improve and consolidate its various main airfields and accommodation, re-grassing Risalpur's landing ground in mid-1925, and, by mid-October that year, declaring Kohat fit for permanent RAF residence. Accordingly, the six squadrons were located as under by the end of October 1925:

Squadron	*Base*	*Commanding Officer*
No 1 (India) Wing		
20	Peshawar	Sqn Ldr J B Cole-Hamilton
60	Kohat	Sqn Ldr T F Hazell, DSO,MC,DFC

* The North-West Frontier had been nicknamed 'The Grim' by the army since the early 1800s.

Squadron	*Base*	*Commanding Officer*
No 2 (India) Wing		
5	Risalpur	Sqn Ldr A J Capel, DSO,DFC
27	Risalpur	Sqn Ldr L M Bailey, AFC
No 3 (India) Wing		
28	Quetta	Sqn Ldr C S Wynne-Eyton, DSO
31	Ambala	Sqn Ldr H S Powell, MC

The year 1926 proved equally peaceful in the context of actual operations, and the squadrons spent much of the year undertaking long-distance formation flights, e.g. Risalpur to Calcutta, practice mobilisation camps, detachments to the forward landing ground at Miranshah, and inter-squadron bombing, gunnery, and sports competitions, apart from a busy purely social calendar.

By December 1926 news reached the squadrons that a grand Air Display was to be held at Delhi early in 1927 before a distinguished audience including the Viceroy, senior Indian and army officials, and other VIPs. Part-purpose for this Indian 'Hendon' was said to be an 'imminent' increase in the establishment of the RAF in India by two further squadrons – shades of John Salmond's 1922 Report's recommendation – and to help officials decide on the types of aircraft best suited for future air operational requirements; though undoubtedly a major reason for such a display was simply to awe Indian spectators with an impressive array of air 'power'.

Rehearsals on the participating squadrons commenced in December 1926 and, at Risalpur, very nearly wiped out the bulk of two squadrons! On 22 December, Nos 27 and 60 Squadrons undertook a simultaneous take-off in squadron strength, with each unit's commander at the apex of his aircraft. Within seconds of the start of take-off it was obvious to horrified spectators that the two formations were converging on a collision course! The unflappable Lionel Bailey, leading his 27 Squadron, characteristically pressed on with his take-off, while Hazell and his 60 Squadron crews on seeing the

FLYING ELEPHANTS. Officers of No.27 Squadron in later 1920s. Back row: ?; Martin; Butcher; Mangles; ?; McKee. Seated: Fressanges; Brodie; Sqn Ldr L M Bailey, AFC (OC); Prendergast; ?. McKee and Fressanges both rose to Air Marshal and were both knighted eventually.

Full line-up of 20 Squadron's Bristol F2Bs, 1929.

impending disaster promptly veered and banked away in any direction which offered escape, scattering in all directions. As pure luck would have it no aircraft or human suffered damage – but a few 60 Squadron crews had noticeably paler suntans thereafter …

The actual Display took place on the Delhi Race Course in 21 February 1927, with all six squadrons giving a series of immaculate formation and other demonstrations worthy of comparison with any contemporary Hendon Air Display.

In March 1927 Air Vice-Marshal Sir Geoffrey Salmond arrived in India to take up his appointment as the latest AOC, RAF India, in which post he was to serve for four years until being succeeded as AOC in February 1931 by Air Vice-Marshal Sir John Steel. Geoffrey Salmond, a pilot since 1913, had seen wide operational command experience during and since the 1914-18 war, particularly in the Middle East zones, and had been something of a pioneer of extending embryo air routes from England and Egypt to India. His experience, allied to a fervent belief in the efficacy of air power, was to stand him – and the RAF in India – in good stead during his tour of duty as AOC, RAF India; contributing much to the gradual elevation in status *vis-à-vis* the army, of the air arm.

He arrived in India during a period described in the *Year Book, 1929* as, 'one of almost absolute peace.' (referring to the years 1927-28) although 'absolute' was too strong a description. Admittedly, tribal insurrections were not on any significant scale during those years on the Frontier, but the period was not without minor incidents which required the RAF squadrons to maintain a constant readiness for possible operations.

Meanwhile, there were such tasks as protection of the regular army relief columns up to such outposts as Chitral and Gilgit, south of the Hindu Kush mountains, between June and October each year; demonstration flights over particular tribal centres where there might have occurred signs of potential unrest; delivery of mail to the many near-inaccessible army Scout posts in the mountains; and sundry one-off flights at the behest of local Political Agents wishing to see their territorial responsibilities from the air. In addition the seemingly unending programme of aerial reconnaissance and photo-mapping continued apace; while several units began practising night-flying, utilising crude flare-paths comprised of spaced goose-neck paraffin flares bordering the designated 'runway' – a practice, it might be added, *still* in RAF use in England during the early months of World War Two!

Just prior to Geoffrey Salmond's arrival another noted figure appeared on the Indian scene, T.E. Lawrence – 'Lawrence of Arabia' – shallowly disguised as Aircraftman T.E. Shaw*, who was aboard the troopship SS *Derbyshire* when it docked at Karachi in the first week of January 1927. A wealth of literature has been (and continues to be) published about Lawrence's life† but in the context of this book it is only necessary to briefly note his RAF service in India.

Posted initially as an Aircraft Hand, General Duties (ACHGD) to the Karachi RAF Depot at Drigh Road, he remained there for over a year before his presence had become something of an embarrassment to certain senior officers at the depot. Requesting a posting to a firstline squadron, Shaw next went to No 20 Squadron at Peshawar in May 1928 and by August was with the unit detachment at Miranshah. By the close of 1928, with troubles brewing across the nearby Afghan border, Shaw's position was – quite falsely – considered by certain Press and political figures to be merely a cover for political intrigue, resulting in his Indian sojourn being abruptly ended, and he was quickly posted back to England, leaving abroad the SS *Rajputana* on 12 January 1929.

Even as the international protest about Shaw's presence so close to the Afghan border boiled up, a number of changes in the overall situation were initiated. For the RAF in India came final confirmation that the two extra squadrons – promised for the past six years – were about to be despatched from England. With this increase in strength in mind, the establishment of the RAF in India was revised and reorganised in October 1928 as follows:

* Lawrence changed his surname to Shaw by deed poll.
† See, in particular, *Solitary in the Ranks* by M. Hyde; Constable, 1977.

No 1 (India) Group HQ		
20 Squadron	Bristol F2B	Peshawar
No 1 (India) Wing		
27 Squadron	DH9A	Kohat
60 Squadron	DH9A	Kohat
No 2 (India) Wing		
11 Squadron	Wapiti IIA	Risalpur
39 Squadron	Wapiti IIA	Risalpur
No 3 (India) Wing		
5 Squadron	Bristol F2B	Quetta
31 Squadron	Bristol F2B	Quetta
Air HQ		
28 Squadron	Bristol F2B	Ambala
Aircraft Park	–	Lahore
Aircraft Depot	–	Drigh Rd, Karachi

This reshuffle of the six resident squadrons took place by January 1929, while the two ex-UK units, Nos 11 and 39 Squadrons, began arriving in stages by mid-January. Initially, however, both of these latest units were just personnel without aircraft, their Wapitis having been crated and shipped separately. In 39 Squadron's case, for example, it only began receiving its Wapitis at Risalpur in early March 1929 and was not up to full 12-aircraft firstline strength until six weeks later.

At that same period several 'new' items of equipment were tested or came into general use. In October 1928 Flight Lieutenant John Baker* of 60 Squadron carried out test-flights of airborne oxygen equipment; an air crew aid first employed by several air services during the 1914-18 war but something of an innovation in India. In December 1928 came the first general issue of parachutes to all flying personnel in India, consisting of Irvin seat packs for pilots and the lap type for rear-cockpit gunners. Again, parachutes had been in use in some air forces, including the RAF in the UK, for several years but their arrival in India was a novelty – and one not particularly welcome to some pilots who objected to the extra

* Later Air Chief Marshal Sir John Baker, GBE,KCB,MC,DFC.

'unnecessary' (*sic*) weight which might affect aircraft performance. Nevertheless, regulations making the carriage of parachutes mandatory soon followed.

Other equipment introduced for more efficient co-operation with the army included the use of W/T and R/T sets during 1927-28, fitted in the rear cockpits of (mainly) Bristol F2Bs. While W/T, using the Morse Code, proved quite successful, R/T procedures proved less effective, although a demonstration of R/T during the Delhi Air Display in February 1927 had been successful. Apart from the technical difficulties of airborne R/T – the F2B airframe was never really satisfactory for safe bonding – the air gunners found contemporary R/T sets too bulky and cumbersome to operate. It should be noted that at that period only two official categories of air crew existed, i.e. pilot and observer. Air gunners in the main were volunteer ground tradesmen on a part-time basis with no advancement in promotion for the task, although a few pence was added to their daily trade pay if 'regularly' engaged in such operational duties. Nevertheless, there was never a shortage of volunteers for the job.

The airborne duties of the back-seat crew men were well described by Air Commodore Tyndall-Carill-Worsley who served as a junior officer-pilot with 20 Squadron flying Bristol F2Bs:

> In the Bristol Fighters and later Westland Wapitis which we flew the Lewis guns were on Scarff mountings and consequently, particularly in the Brisfit (F2B), the aircraftman in the back seat was particularly uncomfortable and had many long, boring, monotonous hours on reconnaissance flights, just hoping the pilot might have a message to pass which would at least break the tedium and keep him in practice.
>
> The accepted method of passing the time was to play noughts and crosses by wireless with the duty operator back at squadron headquarters. This was carried on in a sort of W/T shorthand code which all W/T operators knew, but which few pilots could ever understand. On one of my first

flights, at the end of a long artillery-spotting exercise, I was mystified to be handed a chit from the rear cockpit which was a message from base consisting entirely of four letters. PUFO. It had to be explained to me that this was the traditional way for base to pass permission to return, and was simply an abbreviation for 'Pack U and F---Off!' In between times, when one had permission to bomb, the poor air gunner just had to sit there and take it whilst one went through all the violent manouevres involved in a good beat-up.

One occupational hazard for pilots in those times, especially new ones, was the danger of getting a good wallop on the head from the Lewis gun as the gunner decided to move it from side to side, via a forward arc instead of towards the rear. When doing high-level, as opposed to dive-bombing, the air gunner did the bomb-aiming, using the High Altitude Drift sight clamped alongside the fuselage. To do this he had to lean out over the side and, since there was no intercom, he 'requested' alterations of course by reaching over and thumping the pilot on the appropriate shoulder – the only legitimate opportunity for an airman, except in certain sporting activities, to 'bash' his superior officer ... !

Even as the existing six squadrons in India were being re-organised in location in anticipation of the arrival of Nos 11 and 39 Squadrons in late 1928, the relative calm along the Afghan border was dissipated by events within Afghanistan which had serious repercussions for the British Legation at Kabul, located some two and a half miles west of Kabul city. King Ammanulla of Afghanistan, disturbed by the patent backwardness of his country and people, had begun to institute certain reforms of long-held traditions and customs among his subjects.

Having made a seven-months tour of European countries to see for himself the 'modern' ways of 'civilisation', Ammanulla returned to Afghanistan and proceeded to impose such radical reforms as establishment of the first Afghan Parliament, to be *elected* by votes of all literate Afghans, an increase from two to three years in compulsory army service, new additional taxes to

Scarff Ring and Lewis machine gun mounting in rear cockpit of a 27 Squadron Westland Wapiti. Note pilot's Very Signal Pistol in side 'holster' (bottom rt).

Vickers Victoria, loaded with civilian passengers, prepares to take off from Kabul airfield during the Kabul Legations' evacuations.

cover purchase of armaments, education for Afghan girls along European lines, fixing the marriageable age of girls at 18 and men at 22 years, a decree abolishing the strict *purdah* of all females, and even such trivial decrees as the compulsory wearing of European-style dress (including top hats or other European headgear) for all inhabitants and visitors to Kabul city. He also advocated the abolition of polygamy, ordering that any government official taking a second wife would have to resign his post.

All such decrees were offensive to his Moslem subjects, but they were especially incensed by the attempt to abolish the female *purdah* – involving females no longer being required to wear a veil over their lower faces at all times in public – and the proposed education of girls in the European mode. These latter propositions defied the deeply-held religious beliefs of Ammanulla's subjects, and the many *mullahs* united in protest, openly denouncing Ammanulla as a *Kafir* (Infidel or Unbeliever).

These various decrees by Ammanulla were, in fact, merely the latest examples of his attempts to reform his kingdom over the previous six years. Thwarted in those earlier plans for modernisation, Ammanulla had steadily alienated large sections of his citizenry by measures against certain *mullahs* and tribal leaders (including executions), and disaffected much of his army by drastic reductions in pay and facilities.

Ammanulla forcibly expressed his determination to impose his new proposals in a lengthy series of speeches in September-October 1928 to his 'parliament' despite the overt and unanimous opposition to his wishes by all assembled in his audiences. First to actually openly rebel against Ammanulla were the Shinwaris (literally 'Green Lords'), a fanatical tribe of deeply religious beliefs, who were ordered to adopt European dress, to send a quota of their daughters to Kabul for education, and to pay taxes (which they had never done before). The tribe rose as one man, willing to forfeit their lives if needed, and first attacked Jalalabad, cutting off that city's water supply and closing the Peshawar-Kabul road. Ammanulla's response was to send some aircraft (piloted by Russians) to bomb the rebels.*

* The Afghan Air Force had some vintage DH9s, piloted usually by Russian refugee pilots.

This use of foreign *infidels* by the 'King of Islam' to subjugate faithful Moslems only exacerbated the situation and led to immediate revolts by other tribes such as the Afghan Mohmand who joined forces with the Shinwaris.

Further north an opportunist leader of a band of robbers, named Bacha-i-Saqao ('Babe of the water-carrier' literally), or Habibulla as he became more widely known, seized his chance for greater glory (and possible loot) by joining with the rebel forces against Ammanulla in Kabul. Part-encouraged by a *mullah*'s prophecy that he would one day be king, Habibulla gathered a force in excess of 3,000 disaffected tribesmen and finally made a surprise attack on Kabul on 14 December 1928, capturing some forts one miles north-west of the city and then advancing to the Asmai Heights, west and south-west of the Britsh Legation where his forces were temporarily checked by Ammanulla's dwindling army, then continued their advance on Kabul, passing the Legation en route. As the horde of tribesmen began nearing the British Legation the Minister there, Sir Francis Humphrys, ordered the Legation gates closed and then met Habibulla at the main gate as he arrived who assured Humphrys that the many white people within the Legation would not be harmed.

Humphrys, an ex-RAF pilot himself, had already anticipated possible danger to the Legation and its inhabitants when the first rumblings of rebellion against Ammanulla's reforms were evident weeks before the actual uprising, and on 3 December had approached Geoffrey Salmond with a request for an air mail service to Kabul to maintain communications, and with plans for a possible evacuation of the Kabul Legation's personnel, necessarily by air.

Although in immediate agreement to Humphrys' proposals, Salmond faced a practical problem in that the only aircraft immediately available to him based in India even remotely suitable for an air evacuation were 24 DH9As of 27 and 60 Squadrons and two Wapitis. The nearest aircraft *designed* for carrying passengers in bulk were ten Vickers Victorias of 70 Squadron, based at Hinaidi, Iraq – some 2800 miles away – plus a Handley Page Hinaidi, J7745, at Baghdad which had been detached specifically to convey Sir Denys Bray, the Foreign Secretary to India on 14 December. Salmond promptly requested that the HP Hinaidi be returned but this aircraft was delayed for an engine change, and in its place a Victoria, skippered by Squadron Leader R.S. Maxwell,* was flown to India, arriving at Karachi on 17 December and proceeding to Quetta next day. Tested there in local conditions, the Victoria then moved to Risalpur, ready for evacuation duties, on 19 December.

Meanwhile the Kabul Legation and its crowded residents were in a perilous position. Situated directly between the rebel army and Kabul city, the Legation was surrounded by wild tribesmen whose weapon discipline was non-existent, resulting in many randomly fired bullets riddling walls and windows of the main buildings. All normal communications between Humphrys and the British in India had now been nullified after a final, interrupted wireless message from Humphrys on 16 December, requesting urgent evacuation of all women and children within the Legation.

On 18 December a DH9A of 27 Squadron, crewed by Flying Officer C.W.L.Trusk and Leading Aircraftman G. Donaldson, flew to Kabul carrying a Popham Panel to drop into the Legation grounds for future ground signalling to aircraft, but as the Ninak swept low over the building it was the target of several tribal riflemen and was hit in the engine sump, forcing Trusk to land on the adjacent Afghan Air Force landing ground of Sherpur. As with all the DH9As of 27 and 60 Squadrons detailed for possible use in the evacuation, Trusk's Ninak had been previously stripped of all armament except for a Very signal pistol – in order not to provoke the Afghans in the event of any forced landing – while their crews had been hastily issued with passport visas stamped as 'Visitors'.

Eventually Trusk and Donaldson managed to get inside the Legation, running a gauntlet between the firing of the opposing armies en route, having brought the Ninak's wing generator with them with which they were able to rig up a

* Later, Group Captain, MC,DFC,AFC.

wireless transmitter and receiver and establish a shaky link with Peshawar and Miranshah. Further DH9As flew over the Legation during the next four days but these were told not to attempt any landings as yet, having to be content with parachuting into the Legation a wireless set and sundry items.

Besieged by the opposing forces, constantly under rifle and shell fire from ill-aimed gunners from both sides, and never sure when the Legation building might be actually invaded and a possible massacre perpetrated, the inhabitants of the Legation received at least a temporary respite on 22 December when Ammanulla's troops managed to force the rebel force to retreat north-west, thereby making it possible for Humphrys to contact Kabul city and regain clear communication with the Indian authorities. Accordingly, he immediately sent a message requesting full evacuation to commence next day, using the nearby Sherpur airfield which was now in the hands of Ammanulla's forces. The RAF's response was to despatch a Wapiti, the Victoria, and three DH9As very early on 23 December to Sherpur where they were loaded with women and children in the Victoria and baggage in the other aircraft. This first evacuation flight then returned to Peshawar, and the same process was repeated on 24 and 26 December. By 27 December the Hinaidi transport aircraft was declared serviceable and on the 29th two more Victorias arrived from Iraq. By 1 January 1929 a total of 132 people and baggage had been airlifted back to India.

Within a week, however, Habibulla's tribesmen had gained a victory over Ammanulla's troops and were once more investing Kabul, including regaining control of Sherpur airfield. Then, on 14 January, Ammanulla abdicated and fled the scene by car towards Kandahar, renouncing his throne in favour of his elder brother, Inayatullah Khan. The latter, closely besieged in his own palace, with his garrison force outnumbered by nearly four to one, reigned for merely three days and on 17 January, on condition that he, his family and a small party of faithful officials might be evacuated by RAF aircraft, Inayatullah handed over the Amirate of Afghanistan to the rebel leader Habibulla (who had already styled himself Amir).

The condition was agreed and on 18-19 January Inayatullah and his party were airlifted to Peshawar from where they proceeded by train and cars to join Ammanulla at Kandahar. This royal airlift had been accomplished by two Victorias, piloted by Squadron Leader Maxwell (whose passengers were the ex-King's harem ...) and Flight Lieutenant R. Ivelaw-Chapman* who took aboard Inayatullah and the other male members of his family.

Although Habibulla was now installed as Amir the probable length of his reign seemed to Humphrys to be in immediate question, and he therefore sought permission from the new 'King' to use Sherpur airfield for complete evacuation of all remaining British and other foreign Legations' personnel. This permission was readily granted by Habibulla, who had at no time displayed any personal animosity to the British, even to the extent of providing an airfield guard to ensure safety of the passengers and aircraft.

By then weather conditions had deteriorated rapidly as the onset of the coldest winter for decades took an icy grip, with temperatures plunging well below freezing point and heavy snowfalls. Nevertheless, the DH9A and Victoria crews continued their doughty efforts to evacuate Legation staffs, despite having to endure in their open cockpits dawn take-off temperatures as low as minus-20 degrees Centigrade on occasion, and a constant struggle to attain a 6,000 feet altitude by use of full throttle; the lowest height required to allow them to fly through the mountain ranges peaking at 10-15,000 feet en route.

Geoffrey Salmond had already requested further Victoria aircraft from Iraq to support the three machines already engaged in the Kabul airlift, and five Victorias eventually reached India on 8 February (two aircraft), the 11th (one), the 14th (one), and the 15th (one). Before their appearance the final phase of the evacuation had begun from 28-29 January in the teeth of worsening wintry conditions and, no less importantly, against a background of burgeoning unrest in

* Later Air Chief Marshal Sir Ronald, GCB,KBE,DFC,AFC.

Handley Page Hinaidi, J7745 at Peshawar on return from Kabul, January 1929. Originally built as an HP 24 Hyderabad, J7745 was modified to become the prototype HP Hinaidi (HP 33).

Inayatullah Khan leaving Peshawar by car (seated rear left) on being evacuated by air from Kabul, 19 January 1929.

and around Kabul among tribes unwilling to accept Habibulla as their lawful king.

During the later stages of the airlift Sherpur airfield lay under seventeen inches of frozen snow, frustrating the use of the heavier aircraft, but Humphrys 'mobilised' the local population, aided by every available camel and even a few elephants, to literally stamp out a smooth runway on the impacted ice which permitted four Victorias from Risalpur to land on 24 February and embark yet more passengers and baggage.

On the morning of 25 February 1929 came the final sorties, when seven Victorias, accompanied by the sole Hinaidi, left Risalpur from 0745 hrs and made their way to Sherpur. Landing there, with engines kept running, and within earshot of the sounds of increasing gunfire and turmoil within Kabul city as Habibulla's opponents began their bid to dethrone him, the aircraft took aboard the remaining Legation personnel; the last of these to climb aboard being Flying Officer Trusk, Leading Aircraftman Donaldson, and Francis Humphrys who brought with him the Union Jack, which had flown so long above the Legation buildings, tucked under his arm.

As the last aircraft got airborne its passengers could see flames beginning to engulf the centre of Kabul city. They were the last of a total of 586 men, women and children evacuated overall, apart from some 24,000lb of baggage *et al*; all within a total of 84 active sorties covering a mileage during evacuations alone of 28,160. The only loss had been Victoria J7926 piloted by Flight Lieutenant Ivelaw-Chapman and Flying Officer A.R.S. Davies on 29 January which experienced engine failure and forcelanded with no serious injuries to its crew among a horde of tribesmen, whose captives they remained until 18 February.*

The outstanding success of the Kabul airlift – an entirely mercy mission with no overtones of warlike operations – was hailed by many countries apart from Britain; of the 586 people actually airlifted only 23 had been of British nationality, 344

* See *Wings over Kabul* by A. Baker/ACM Sir R. Ivelaw-Chapman; Kimber 1975 for full personal account by Ivelaw-Chapman.

'British Indian', and the rest of eleven different foreign citizenships. Recognition of the RAF's splendid achievements took the form of the awards of Air Force Crosses (AFCs) to Squadron Leader Maxwell, Flight Lieutenants D.F. Anderson and Ivelaw-Chapman, and Flying Officers Trusk and Anness; while Leading Aircraftman Donaldson received an Air Force Medal (AFM). Sir Francis Humphrys and his wife were later the centre figures of a unique investiture at Buckingham Palace where they were decorated side by side with the KCMG and DBE respectively by HM King George V. As for the self-proclaimed Amir Habibulla Khan II, his reign was brief, being ousted from his Kabul throne by Nadir Khan, a member of the dynastic Duranni family, who then took the crown for himself – despite his previous proclamation of intent to restore Ammanulla as Amir ...

In the aftermath of the Kabul airlift, Geoffrey Salmond's official report included his views of the value of having heavy transport aircraft on his RAF India establishment; pointing to the fact that such squadrons existed already in Iraq (No 70) and in Egypt (No 216). In doing so he was, unknowingly perhaps, echoing the proposals of Squadron Leader E.J. Hodsoll, nearly two years previously in an article 'Air Lines of Communication,' published in the July 1927 issue of *The Journal of the United Service Institution of India*. Hodsoll had also pointed out the lack of large transport aircraft in India compared with the Middle East Commands, and had proposed the provision of at least a Flight of four De Havilland Hercules civil airliners, suitably converted to RAF requirements, for transporting army formations, artillery, and other necessary equipment to relieve and/or reinforce army garrison forts along the North-West Frontier Province, thereby eliminating the need for the traditional thousands of troops normally required to merely protect any land-bound relief columns, apart from the drastic reduction in the time needed for such movements.

Although Hodsoll's views had no immediate effect, the Kabul evacuation appears to have highlighted his prime theme because in 1929 a Handley Page Clive transport joined the HP Hinaidi at Lahore to form the Heavy Transport Flight,

Bristol F2B, J6774, 'C' of 20 Squadron, 1930, an ex-5 Squadron aircraft. The upper wing 'gun' was actually a Hythe camera-gun for air firing practices.

Wapitis J9717 & J9719 of 60 Squadron returning from a raid on Afridi tribesmen, August 1934.

followed by two more Clives in the following year. This unit was retitled the Bomber Transport Flight in 1932 and remained as such until being absorbed by No 31 Squadron with effect from 1 April 1939 when 31 became a bomber transport unit, by which date the main aircraft were Vickers Valentias. Initially the Heavy Transport Flight was little used for pure troop-carrying, becoming more a freight transport unit, thereby providing extra support to the Wapiti squadrons already well-used to parachute supplies at pre-arranged forward posts for use of army columns.

Only in the 1930s were the large transports to come into their own as a recognised boon to troop movements, particularly as adjuncts to the regular relief of such forward posts as the 15-17,000-strong army force at Chitral, but equally importantly as rapid resupply vehicles to army forces engaged in tribal operations. It calls for little imagination to realise the difficulties of flying heavily-loaded, cumbersome bomber-transport aircraft over the treacherous mountain terrain in searing temperatures, let alone delivering their loads accurately, constantly countering the extremes of turbulence over and between the mountain peaks.

Once the Kabul evacuation had been completed, the squadrons returned to more routine duties for much of 1929, although in that year were born the seeds of further rebellions, part-prompted by the civil disobedience campaign inspired by Mahatma Gandhi via the Indian National Congress, which was to spread into open revolt from 1930. For the RAF squadrons in India 1930 saw the start of re-equipment for those units still flying the World War One vintage DH9A. No 60 Squadron received its first Wapiti IIA on 8 March 1930 and completed full conversion to their new aircraft by mid-May; while No 27 Squadron commenced re-equipment with Wapitis in April 1930.

Although similar in size – the original Wapiti design had incorporated many standard DH9A components – the Wapiti differed greatly from the Ninak in many respects. In place of the DH9A's wooden wing structure the Wapiti had a metal frame, while the change from the Ninak's well-worn 400hp

Normally a Wapiti was 'started up' via crank handles each side of the nose engine compartment, but the traditional three-man muscle-power was often necessary; illustrated here on a 27 Squadron Wapiti in 1937.

Vickers Valentia, K3167 of No.216 Squadron over Lake Victoria in April 1937.

Liberty engine to the Wapiti's 550hp nine-cylinder, radial, air-cooled Jupiter powerplant offered a far superior performance range. Although rather heavier on the controls than a Ninak, the Wapiti was somewhat easier to fly, being fully aerobatic and totally stable. Its take-off and rate of climb – 1140 ft/min ideally – was also a huge asset when lifting a full load off the high altitude landing grounds of the North-West Frontier. If pilots welcomed the Wapiti, rear-seat gunners found little improvement in their 'office'. Still encircled by a Scarff Ring with its single Lewis machine gun, the gunner still had no form of seat for any lengthy flights, but there was provided a detachable control column (although no duplicate rudder controls) which could be connected in the event of the pilot being incapacitated, though no few airmen were given the opportunity thereby to take over normal straight and level flying if their pilot was amenable.

The use of that duplicate 'joystick' was quickly demonstrated when, on 17 May 1930, a Wapiti from 11 Squadron, crewed by Flying Officer Richard Stroud and AC1 Charles Samuel Wiltshire, was engaged in low-bombing a gathering of tribesmen near the Khyber Pass. At 600 feet the Wapiti came under heavy rifle fire and one bullet hit Stroud in the neck. Signalling to Wiltshire to take control Stroud then collapsed. The gunner immediately slotted in his control column and eased the aircraft away from the target. He now faced two grim alternatives. Either take to his parachute and face the possibilities of torture or death at the hands of the tribesmen, or attempt to fly the Wapiti back to base at Risalpur. Wiltshire had no knowledge of piloting, and in any case could not be certain whether Stroud was dead or merely unconscious.

He chose to try for Risalpur. Arriving over the landing ground, Wiltshire attempted to set the Wapiti down but landed too fast and crashed badly, sustaining serious injuries to which he succumbed in hospital the same night. Wiltshire's courageous act was officially recognised by a near-unique posthumous promotion to the rank of Corporal – an 'award' which may seem curious to any present-day airman but indicative of the esteem and authority enjoyed by a junior non-commissioned officer in that period.

The Wapiti was to bear the brunt of operations in India for almost a decade, replacing Bristol F2Bs on other squadrons by early 1932 and proving itself ubiquitous in several roles. Its prime operational roles continued to be bomber and/or army co-operation, but it also undertook regular long-distance flights. For this latter purpose an extra fuel tank was normally fitted between the legs of the undercarriage, from which fuel was pumped by a wind-driven auxiliary pump located on the right side of the pilot's seat. One of the few flying vices of the Wapiti was its permanent tendency to swing strongly on take-off and, especially, during landing – 'Stick right back and fully stalled if you want a neat three-pointer' (*sic*) – so that, with no breaking system, Wapiti landings by inexperienced pilots were all too often expensive.

CHAPTER TEN

Red Shirts and Fakirs

After the near-peaceful years 1927-28, with the exception of the Kabul operations, 1930 brought a resumption of intense activities involving all squadrons. Gandhi's civil disobedience movement instigated in 1929 had received active assistance within the North-West Frontier Province from one Abdul Ghaffar Khan of Utmanzai (some twenty miles north-east of Peshawar) who commenced organising the Frontier Youth League in and around the Peshawar district. Politically active since 1919, Abdul Ghaffar Khan joined Gandhi's Congress Party in 1920 and by 1929 had begun forming his Youth League from dissident Pathans into an unarmed but para-military force which used an elementary form of distinctive uniform of shirts dyed with brick-dust, thereby earning the generic title of 'Red Shirts'. These young militants trained along military lines quite openly, yet were virtually ignored by the Indian government for months, and the League gained strength in numbers rapidly as an anti-British force.

Recruiting for his Red Shirts all along the Frontier, on 23 April 1930 Abdul Ghaffar Khan held a huge rally in Peshawar and incited anti-government hatred among his audience, resulting in his arrest along with nine other leading figures in the movement. These arrests provoked savage rioting and anarchy for the following eight days which were only subdued by the army taking control of the city eventually.

Meanwhile agitation spread in waves among surrounding tribes, particularly the Afridis, leading to an increase in recruits to the Red Shirts. Rebellious speeches were made openly in Kohat native city which ended with the arrest of six Congress leaders; while in Waziristan the Mhadda Khel tribe invested the Dhatta Khel Tochi Scout post on 11 May. General unrest had

Wapiti, J9742 of 27 Squadron, loaded with four 112lb RL and four 25lb Cooper bombs, about to leave Miranshah for a raid on rebel Mahsuds, July 1930.

Wapitis K1309 & K 1260 of 5 Squadron.

Wapiti's under-wing racks loaded with 25lb Cooper bombs (31 Squadron). Note pilot's Vickers machine gun on fuselage.

now spread through the Vale of Peshawar where Red Shirts sent supplies and arms to rebel Mohmands then extended their influence north to Bajaur and through the Tirah area of the Afridi, south-west of Peshawar. From mid-May the RAF was heavily involved in operations against the various centres of unrest, mainly attempting to disperse any *lashkar* (large force) of tribesmen gathering.

One such force of an estimated 7,000 Afridis at Tendai was thoroughly bombed and strafed in the Tirah area, while another *lashkar* of 2,000 rebel Red Shirts advanced on Peshawar where battle was joined for 48 hours. Suffering heavy casualties this last force finally retreated across the Khajuri Plain on 6 June, pursued and harried constantly by aircraft.

One casualty linked with these operations occurred on 13 June when a Wapiti from 60 Squadron, crewed by Sergeant E.A. Hindelang and LAC W.H. Lansdell, took off from Kohat fully loaded with bombs and ammunition, only to stall and crash at the edge of the airfield, then erupt in flames. A nearby airman, Leading Aircraftman Robert E. Douglas, ran to the burning wreck and found Lansdell lying some six feet from the aircraft, his clothes well alight. Quenching the flames with a hand extinguisher, Douglas dragged the badly burned air gunner clear of the burning Wapiti and got him into an ambulance. Then, despite being well aware that the burning aircraft had unexploded bombs aboard, Douglas went back to try to extricate the pilot and was within twelve yards when a bomb exploded. Realising there was now no hope of retrieving the pilot alive, Douglas started to get clear but was only 30 yards away when a second bomb exploded. Douglas's unselfish bravery was acknowledged by the award of a British Empire Medal (BEM) in March 1931.

July began with a Mahsud uprising in Waziristan where some 3,000 tribesmen invested the South Waziristan Scout post on 6 July. Bombed and strafed daily from the air for the next six weeks, these Mahsuds eventually scattered to the hills and their leaders were imprisoned. Only the Badinzai Mahsuds remained hostile and these held out against continuous bombing until their final surrender on 27 July.

Throughout August hostile raids on outlying army posts continued, while on 7 August a *lashkar* of 3,000 Afridis attacked Peshawar again. The increasingly grave situation there led to martial law being declared in the Peshawar district on 15 August and by the end of that month this major uprising petered out, although minor sporadic troubles were to continue through the oncoming winter months. Elsewhere widespread actions against rebelling tribes took place during 1930-31 calling for intense active operations by the RAF squadrons, and were exacerbated in March 1931 when the Red Shirt leader, Abdul Ghaffar Khan, was released from prison and immediately attempted to ferment further agitation by reintroducing the civil disobedience movement. His efforts only partially succeeded and by 3 October hostilities with the main Afridi rebels were concluded. Abdul Ghaffar Khan was then re-arrested on 24 December and immediately deported with four of his chief henchmen.

For the moment, at least, the Red Shirt campaign had collapsed. In its wake several RAF men received awards, including an OBE to Squadron Leader R.G.H. Neville, MC,* OC 60 Squadron and a unique double DFM award to the 60 Squadron air crew Sergeant R.A. Falconer (pilot) and his usual air gunner Leading Aircraftman J.A. Dwyer.

These Red Shirt operations proved to be the swan song of the long-serving Bristol F2B's active firstline service with the RAF. In February 1931 No 5 Squadron began re-equipment with Wapitis and had completed its conversion by May, while No 31 Squadron received its first Wapiti on 13 February 1931 and fully re-equipped by 16 April. In the following September No 28 Squadron began exchanging its F2Bs for Wapitis. These changes left only 20 Squadron soldiering on with F2Bs but in January 1932 it too commenced gradual re-equipment with Wapitis and by April had finally disposed of all F2Bs on unit strength. It flew its ultimate F2B formation flight on 13 March 1932 – the final salute to an aircraft design which had served the RFC and RAF faithfully for fifteen years, virtually

* Later Group Captain, OBE,MC.

Wapitis of 5 Squadron dropping supplies to a forward army column.

Wapiti of 28 Squadron at Delhi, complete with the remains of a vulture ... Note unit badge on nose of aircraft.

Wapitis K1282 & K1278 of 11 Squadron on patrol, ca. 1931.

MATING SEASON. Wapitis J9741, 'W' and K1307, 'N' of 5 Squadron get too friendly, Quetta, February 1934.

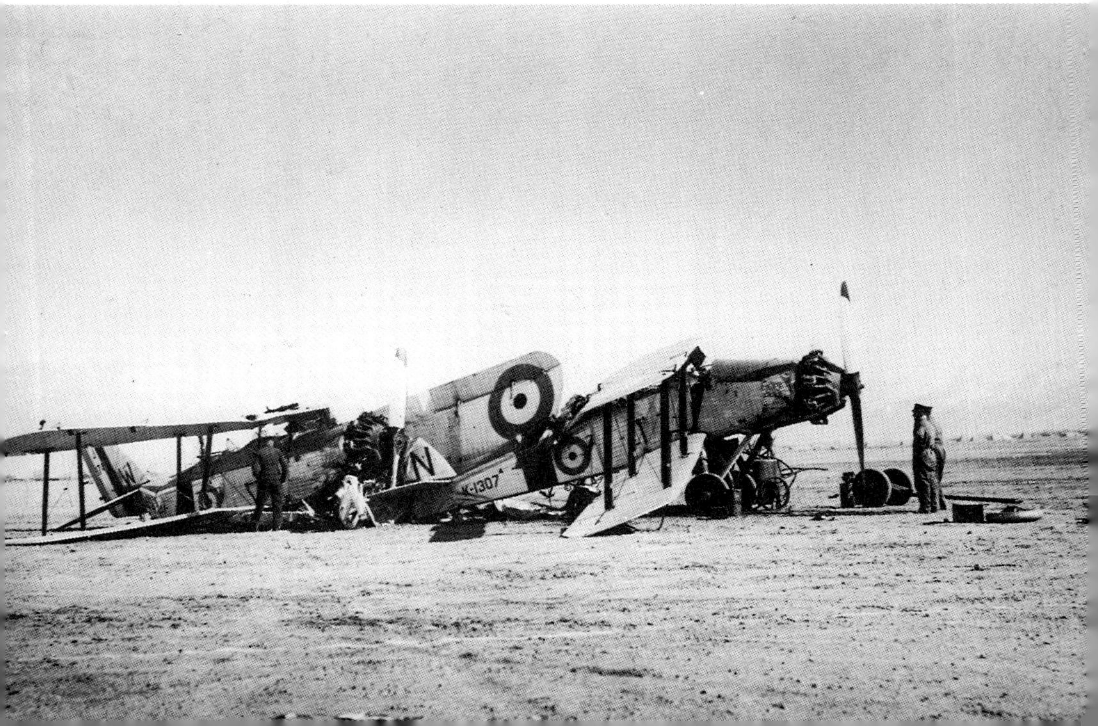

SWAN SONG. The ultimate Bristol F2B flight by No.20 Squadron leaving Peshawar on 13 March 1932, prior to re-equipment with Wapitis.

Wapiti, J9752, 'G' of 20 Squadron setting out on a bombing raid. The long hook under the fuselage was a standard means of retrieving messages from ground troops.

unchanged from its original design – and it was appropriate that 20 Squadron should provide the finale to the career of the 'King of Two-seaters', having been the longest-serving F2B unit, 1917-32. At the same period that 20 Squadron retired its F2Bs, Nos 11 and 39 Squadrons also began converting to new types of aircraft, with both units exchanging their Wapitis for Hawker Harts; No 39 commenced re-equipment in November 1931, with 11 Squadron following suit by February 1932. The Hart offered its crews higher cruising and maximum speeds than the Wapiti, coupled with increased range, though armament and bomb load capacity remained the same.

The early months of 1932 brought a further series of operations against recalcitrant tribes when certain sections of the Red Shirts still at large gave support to an uprising led by the Fakir of Alingar in March, while elsewhere action was taken against the Nawab of Dir's forces in the Milan Kalai area; both insurrections being countered by bombing raids. These were followed in July by an inter-tribal dispute between Upper and Lower Mohmands with the RAF being involved in support of the latter faction. In September came the regular bi-annual relief of Chitral – an outpost army garrison in the eastern hills of the Hindu Kush where army personnel undertook in turns a two-year stint of frontier guard duties – involving a 140 miles route march through some of the wildest tribal territory between Administered Territory and the India-Afghanistan border.

This regular relief operation was, of course, well known to the hills tribesmen who looked upon it as an excellent opportunity for ambush and possible loot. As one RAF officer described this particular operation:

> The Pathans regarded such army operations almost as we viewed inter-unit sporting fixtures, despite the bombing and gunning involved. They could break the monotony of farming and guarding their goat flocks from other tribes by letting off steam against the British. They always loved a fight, particularly when there was a prospect of gaining fresh rifles and ammunition as plunder. In a sense such clashes

> were almost impersonal in that the tribesmen had come to regard the British soldier as a worthy opponent against whom they might test their mettle.

On this occasion Wapitis from 27 and 60 Squadrons were detailed to provide continuous air escort and supply support for the Chitral Relief Column which set out from Dargai on 13 September. Within days the Wapiti crews were necessarily engaged in bombing and strafing sorties against Shamozai Mohmands who had chosen to attack the column, and such operations were to continue until the tribesmen finally withdrew from the column's line of march in mid-October.

Once Chitral had been successfully relieved, air operations resumed more normal routines, with an emphasis on training, familiarisation and reconnaissance patrols along the Frontier. In March 1933, however, the aircraft swung into action yet again, this time in the Tochi area where supporters of the ex-Amir Ammanulla of Afghanistan attempted to stir up the Mahsud and Wazir tribes into an attack on Kabul. Prompt bombing of their villages quickly convinced the two tribes that any such venture would be ill-advised ...

Within weeks two further uprisings occurred in the same area, north-west of Peshawar, where a pretender to the Afghan throne sheltering with the Khan of Kotkai was attempting to raise the tribes to depose Nadir Shah, the incumbent Amir, who immediately requested action by the Indian authorities. Although some tribes submitted to the warning leaflets dropped by air, the Khan of Kotkai remained obdurate.

In consequence offensive air actions began. Starting on 1 August the aircraft of four squadrons, 11, 27, 39 and 60, bombed the Kotkai village – a very small encampment situated high on a steep hillside – to such precision that by 13 August the Khan submitted to government terms and expelled the pretender. As these operations were taking place the age-old feuding between the Lower and Upper Mohmand exploded into furious war. The Lower Mohmand received assistance from the government which planned to build a road into the Gandab Valley extending via Ghalanai to Yusuf Khel, with the

Wapitis of Nos 27 and 28 Squadrons gathered at Kohat on 5 February 1934.

Trio of Wapitis from 27 Squadron (K1283, K1299, K1254) led by Flg Off Johnny Darwen (later Group Captain, DSO,DFC & killed in action 7/10/43 over Italy). Another pilot here was Flg Off Freddy Lambert, later Wg Cdr, DSO,DFC.

Lower Mohmand's concurrence. Aircraft were used to disperse the Upper Mohmand, inflicting nearly 100 casualties killed or wounded at a cost of less than twenty government troops and one aircraft forced down and later burned by tribesmen (although its crew escaped injury and reached some nearby army troops).

In general, air operations against tribal villages harbouring dissidents were conducted along strictly observed lines, these having been laid down by Hugh Trenchard in the early 1920s and adhered to by his successors. First came political persuasion, then plentiful warnings to the tribes of impending air action if 'diplomatic' reasoning failed, followed ultimately by actual bombing. The somewhat light-hearted account of this procedure was contained in an anonymous diary of the late 1920s:

> The tribes found the strain of good behaviour a bit too much at times. There were definite rules to this game strictly observed on both sides. Firstly, the Political Agent (PA) would summon the tribal leaders to a *jirga* (meeting) at which he would tax them with their sins and pronounce government penalty – usually a fine of rifles and/or rupees. This was but the first of several meetings which the culprits used for the successive stages of denying responsibility, suggesting that half – maybe a quarter – of the sum would be more in keeping with their resources, offering so much down and the rest in instalments, and finally flatly refusing to pay.
>
> The PA would then issue his ultimatum which usually gave the tribes another week in which to get their caves in the nearby hills habitable. The day before this ultimatum expired squadrons would demonstrate over the village and drop leaflets tied to small pieces of rock and containing a reminder that all women and children should be removed. The fact that nobody could read was immaterial – everybody knew the rules. The war itself might last anything up to a fortnight before the tribesmen would indicate their willingness to try to raise the necessary (fines). It was

> generally accepted that the surrender was not brought about by the bombing, most of which made little impression on the strongly-built forts and towers, but by the myriad vermin which infested the hill caves!

It should be noted, however, that while the foregoing remarks were generally true of air operations in the 1920s in India, later years were to bring more serious and widespread actions, not all of which could be preceded by the 'normal' procedures depending upon the urgency for immediate operations. Certainly, the layman thought of RAF air crews bombing 'innocent women and children' of the Indian tribes had been a major complaint against Trenchard's air control policy by his political opponents in the United Kingdom for many years, albeit voiced by people (usually) with little or no personal knowledge of actual operational conditions in Iraq, India, *et al.*

In August 1935 serious trouble erupted again in Mohmand territory, fermented by an old outlaw, the Hadji of Turngzai, and his three sons. On 14 August the eldest son, Badshah Gul gathered a *lashkar* of about 2,000 men of various Upper Mohmand tribes with the intention of destroying the recently completed Gandab military road along with its army posts. The RAF dropped the customary warning leaflets of punitive bombing if these tribesmen failed to withdraw, then flew a mass 38-aircraft demonstration flight over the rebels' villages – to no avail.

In consequence, a phased bombing programme commenced on 20 August by 27 and 60 Squadrons and continued into September when night sorties were added to the operations, beginning on 9/10 September. These night raids were only conducted in clear moonlight conditions due to the obvious hazards of flying between mountains uncertainly mapped yet paradoxically proved exhilarating from a pure flying viewpoint by at least one 27 Squadron pilot. Winging through towering rock faces and stygian-black bottomless gorges illuminated sporadically by silver shafts of moonlight, this pilot was inspired to burst into song ... and his astonished air gunner could later claim to have been 'serenaded' at 7,000 feet in the early hours of the morning with extracts from a Gilbert and Sullivan operetta!

On the ground, however, the army was engaged in almost nonstop close fighting throughout September, culminating on 29 September with a pitched hand-to-hand engagement on a high point known as Point 4080, where the British defence commander, Captain G. Meynell of The Guides (12th Frontier Force Regiment) was killed conducting a courageous defence against far superior odds, and was later awarded a posthumous Victoria Cross. The battle for Point 4080 proved to be the finale to these prolonged operations and on 1 October the tribes submitted to terms.

The RAF's contributions to these latest anti-tribal operations had been restricted by several circumstances. Both 27 and 60 Squadrons had only two Flights available for flying, the third Flight in each unit being rested at the Hill Depot during the hot season as normal. The more serious reason was a tragic occurrence at Quetta, home of Nos 5 and 31 Squadrons. At 0300 hours on Friday, 31 May 1935, an earthquake of two main shocks, lasting merely one minute, followed by tremors, struck part of Quetta RAF accommodation and the nearby Quetta native city. Of the RAF buildings all except the aircraft hangars collapsed, married quarters were razed to the ground, and other nearby bungalows suffered severe damage. In the wreckage 55 RAF men (including two officers' children) and 66 Indians were killed, while hundreds of others sustained injuries of varying degrees. In the native city, just south of the military cantonment, 30,000 of a 65,000 population were also killed. Of the two squadrons' aircraft, only three of 31 Squadron were undamaged, while all of 5 Squadron's machines were unserviceable.

Over the following weeks most of the RAF personnel and equipment had been evacuated from the scene of the disaster and it was to be months before the RAF in India could be declared fully re-established at normal strength. The necessary relocation and reorganisation resulting is illustrated by the September 1935 state:

Wapiti K1309 'F' of 5 Squadron, piloted by Flg Off Emerton.

27 Squadron's Wapitis on a raid against the Fakir of Ipi, March 1939.

RAF India HQ, Simla	
Aircraft Depot	Drigh Road, Karachi
Aircraft Park	Lahore
5 Squadron	Risalpur
28 Squadron	Ambala
31 Squadron	Karachi
BT Flight	Lahore
No 1 (India) Group HQ, Peshawar	
20 Squadron	Peshawar
No 1 (India) Wing HQ, Kohat	
27 Squadron	Kohat
60 Squadron	Kohat
No 2 (India) Wing HQ, Risalpur	
11 Squadron	Risalpur
39 Squadron	Risalpur

The stations listed were unit bases, but the RAF continued to maintain several forward landing grounds, such as Razmak and Miranshah, where individual Flights from the squadrons were regularly detached for two months in rotation, thereby maintaining an operationally-ready element close to the Afghan border.

With the tribes returning to a more peaceful co-existence temporarily, 1936 was a year of almost routine duties for all the squadrons. Regular contact flights were maintained with all forward army posts in such tribal territories as Mohmand, Afridi, and Waziristan areas, often involving the crews in four hours' flying per sortie. Air reconnaissance in general, and photography in particular, was the prime role, and ever-conscious of any tribesman's hobby of practising his marksmanship, crews were briefed to fly no lower than 3,000 feet when possible. Such flights, especially during the hot seasons of roughly April-October, meant that flying crews encountered extremes of temperature on most sorties, setting out swaddled in Sidcot flying suit, scarf, thick underwear, and a flood of perspiration, only to be near-frozen at altitude, then return to the baking heat at ground level of an airfield.

Other occupations regarded as normal practices included

periodic training detachments, involving flying to emergency landing grounds with only tented accommodation and bare facilities for aircraft maintenance, and, in 1936, the first so-termed Reinforcement Flights to Singapore by whole squadrons* – pursuant to a new policy whereby India-based units were able to reinforce British forces in the Middle and Far East in the event of a major war. Such a flight to Singapore took seven days, in stages, over largely unmapped jungles and plains with few facilities en route for servicing, refuelling, and accommodating aircraft. No 60 Squadron despatched its twelve Wapitis, accompanied by two transport aircraft from the BT Flight, on 7 February 1936, and ten Wapitis reached Singapore on 13 February.

Of the two non-arrivals, one Wapiti had suffered serious damage from an air collision with a large vulture and had to forceland; while the second loss was a Wapiti which, in low cloud, crashed into a hillside, killing its pilot, 120 miles south-east of Delhi. Its air gunner, Corporal A.C. Wood, though thrown clear and in a dazed condition, plunged into the burning wreckage attempting to extricate his pilot (Flying Officer A.O.D. Cox), receiving severe burns which later cost him amputation of his left hand. Wood's courage was recognised by the award of the Albert Medal in the *London Gazette* dated 22 January 1937.

In 1935 the Tori Khel Wazirs, a particularly fierce tribe, had signed an agreement with the Indian government's representatives giving the government unopposed access to their territory in return for an increase in tribal allowances. In the following year, nevertheless, the Tori Khel were to be among the chief instigators of a sudden revolt which rapidly burgeoned into widespread rebellion among other tribes.

The spark for this latest conflagration was a clash of religions when, in March 1936, a Hindu girl from Bannu eloped with a young Moslem and – in the Moslem custom – was converted to her fiancé's faith preparatory to marriage, taking the name of Islam Bibi. Before the marriage could take place, however, the

* There had been several previous 'Reinforcement Flights' to Singapore but none in full squadron strength.

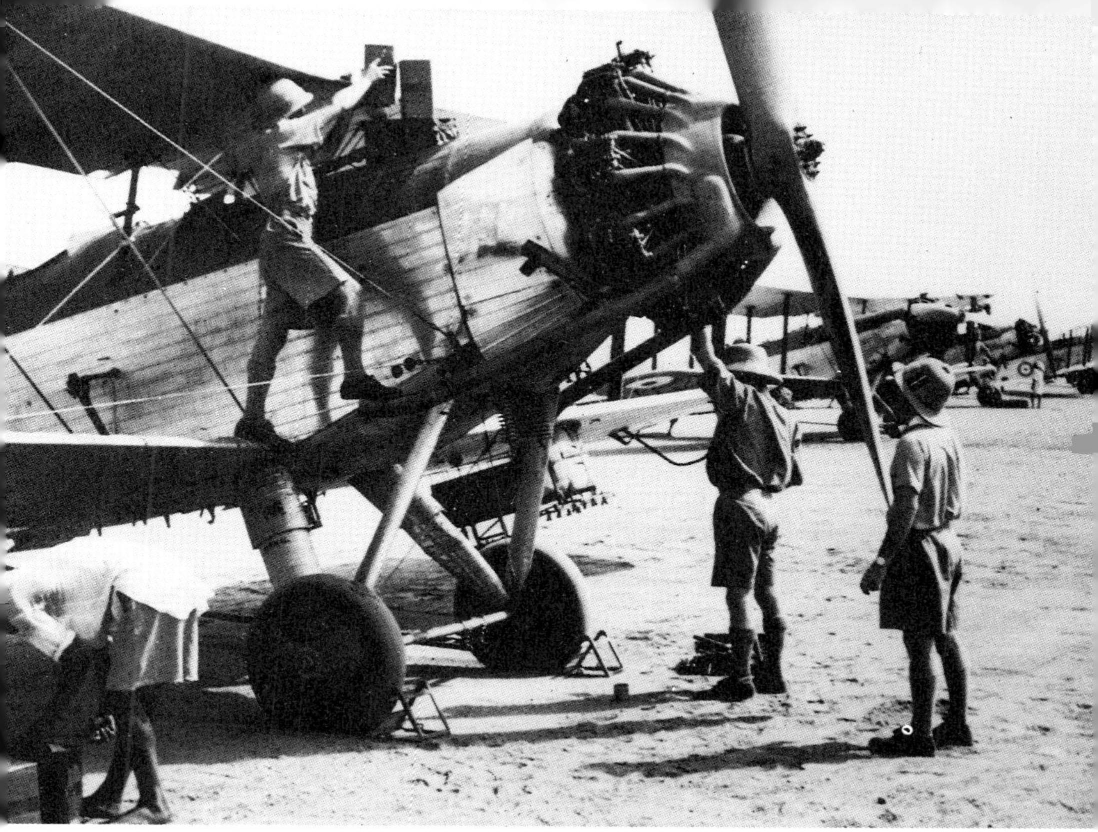

Refuelling a 5 Squadron Wapiti the hard way, Khanpur, 1938.

HIGH COUNTRY. Hawker Harts of Nos 11 & 39 Squadrons over part of the Himalayas at 20,000 ft on 17 October 1932.

British retrieved the girl from Unadministered Territory and returned her to her Hindu parents who brought a charge of abduction against the girl's would-be husband. Since Islam Bibi was now a Moslem this created fury among the Moslem tribesmen, led in protest by one Mirza Ali Khan, the Fakir of Ipi.

Aged about 40, Mirza Ali Khan, a member of a Tori Khel Wazir tribe-clan, had until then spent his life as a deeply religious devotee, taking up permanent residence in the village of Ipi in 1926 (hence his Fakir title) and gaining great respect as a holy man of near-saintly status among the Wazirs and neighbouring tribes. His influence can be judged by the fact that by April 1936 he had persuaded three major tribes, the Tori Khel Wazirs, the Mahsuds, and the Bhittani, to put aside their ages-old traditional differences to form a cohesive anti-British fighting force.

Then, in the autumn of 1936 the Fakir announced himself to be the Champion of Islam and called on all tribes for a *Jehad* (Holy War) against the *feringee* (foreign infidels) – meaning the British and their allies. It was the signal for anti-British insurrections on a wide scale that would last, spasmodically, until after the start of World War Two in India.

The first direct clash with the Fakir's forces came in late November 1936 when two army columns advanced into the Khaisora valley supported by Nos 5 and 27 Squadrons in air supply roles. These army forces combined and were met with fierce opposition, losing 24 killed and over 100 others wounded, but succeeded in ousting the Fakir of Ipi from that area and forcing him into hiding in the hills for the time being.

Within days, however, the Fakir had gathered a force of some 700-800 men from various tribes and was urging others to join his cause. On 23 December he was finally located in a small fortified camp at Arsal Kot, about twenty miles south of Miranshah, and on the edge of Mahsud territory, from where he was attempting to recruit the Mahsuds to join with his Tori Khel force. Though a very small target – approximately 140 yards by 70 yards in perimeter – and situated in particularly remote, mountainous country, it was decided to bomb Arsal

Hawker Hart, K2109 of 11 Squadron with extra fuel tanks.

Hawker Audaxes (nearest K4844) at Miranshah's fuelling (POL) point.

Kot thoroughly as an object lesson to would-be followers of the Fakir. This task was given to 60 Squadron who planned a series of 18 Flight raids spread over three days.

On 31 December these raids commenced and by the next day had accomplished the objective with deadly accurate bomb-aiming, achieving over 60 per cent direct hits with 230lb bombs apart from the near-misses, leaving Arsal Kot a complete ruin. The Fakir had withdrawn to nearby mountain caves during the bombing and remained there, still fermenting unrest, until mid-January 1937, resulting in further air and ground attacks in the general Khaisora area which continued through February, March and April almost without pause. In these operations Nos 5, 11, 20, 27, 39 and 60 Squadrons were all at some period heavily engaged, along wtih the Valentias of the BT Flight; these latter transport aircraft being utilised as bombers on occasion.

By mid-April 1937 the overall army/RAF operations were reaching their peak of intensity, with some 45,000 regular and militia troops engaged. Of the RAF squadrons, 20 Squadrons now operated from Miranshah; 11 and 39 from Arawali; 27, 60 and the BT Flight from Kohat (being refuelled and re-armed at Miranshah). No 5 Squadron had returned to Risalpur by then, and No 28 Squadron was brought forward towards the end of April to Manzai from its usual base at Ambala.

At Miranshah, the main forward operational airfield, at the height of these operations, the constant movement of aircraft coming and going gave the place an atmosphere of near-perpetual motion. Overheated engines bellowed in a nonstop chorus, blasting minor whirlwinds of rock-dust and grit into the blistering sunlight and pasting a coat of sweat-streaked filth upon the long-suffering ground crews toiling to refuel and re-arm each incoming machine. With eyes reduced to mere slits in masks of sand and sunburn, the Erks remained cheerful and ever-ready to offer a word of encouragement to the stiff-legged air crews as these climbed out of their aircraft for a brief respite between sorties. Those air crews were given little rest, averaging four or even five sorties each day if aircraft serviceability permitted, and this

Hawker Hart, K2119 in 39 Squadron markings over the Kabul river.

Vickers Valentias of 31 Squadron at Dum Dum airfield, India in 1936.

intense pace of flying was probably part responsible for several accidents and some fatal casualties.

On 13 April a Wapiti of 27 Squadron (Pilot Officer J.B. Pascoe-Webb and Leading Aircraftman R.W. Gibbs) slewed on take-off for a bombing raid and crashed, its bombs exploding and killing both men; while 5 Squadron lost a pilot killed at Manzai when he crashed and the aircraft blew up. On 18 April Pilot Officer Johnny Darwen of 27 Squadron had his Wapiti's engine fail completely over the besieged Ladha Scout Post, just south of Razmak, but managed to force-land safely, despite still having his bombs aboard. He remained with his aircraft, isolated for weeks before eventually being safely retrieved.*

Tribal attacks on army scout posts, convoys, or other formations continued apace throughout April and the army proceeded to launch a determined effort to 'clear' the Khaisora Valley at least in its efforts to pin down the Fakir of Ipi, supported mainly by 27 and 60 Squadrons which maintained continuous air patrols, interspersed with bombing raids, daily over a wide area, resulting in bullet-riddled wings and fuselages becoming a common sight on returning Wapitis.

By May the Khaisora Valley had to all intents been 'cleared' and the army now advanced towards Arsal Kot, still the refuge of the Fakir and his (mainly) Tori Khel force. The army's advance was again supported by 27 and 60 Squadrons, but now included Valentias of the BT Flight which air-dropped supplies by parachute to the infantry columns; a task rich with hazard among the towering mountains as indicated in the official historian's account: 'The surrounding country was so precipitous and air currents so strong that one lightly loaded parachute instead of dropping on the Plain was carried 3,000 feet *above* the aircraft before finally coming to the ground.'† Nevertheless, the BT crews achieved a total drop of some 13,000lb in 101 loads.

On the night of 19 May the Fakir of Ipi finally left his refuge

* Group Captain J. Darwen, DSO,DFC later achieved near-legendary fame as a fighter leader in the North African campaigns of 1942-43, but was killed in action on 7 October 1943.

† *Official History of Operations*, 1936-37; HMSO.

at Arsal Kot and moved to the shelter of a section of Mahsuds some six miles away. By then his Tori Khel supporters, increasingly bored by the lengthy hostilities, were ready to speak peace to their opponents, and sent representatives to a *jirgah* at Miranshah on 5 June, after which peace terms, including the usual punitive 'fines', were agreed with this particular tribe.

Rejected by the Tori Khel – at least temporarily – the Fakir of Ipi next assembled a force from the Bhittanis, Mahsuds and some Afghans from across the border. Of these the Bhittanis were the first openly to continue the rebellion, attacking army posts, other tribes in British-Administered Territory, and generally seizing loot wherever they could. For the first three weeks in June 1937, Nos 27 and 60 Squadrons maintained a bombing offensive against Bhittani villages and centres with considerable success. By then the overworked BT Flight had been reinforced by the arrival of four more Valentias of 70 Squadron from Iraq on 31 May.

By 1 July the Indian government sanctioned the RAF to bomb *any* village suspected of harbouring the Fakir of Ipi or his followers, and on 8-10 July, 27 Squadron bombed Razin with 230lb bombs successfully despite having to fly very close to the adjacent 12,000 feet high sheer mountain range on their run-in. As one ex-airman serving on 27 Squadron remembered this particular operation:

> My pilot for these ops was an ex-UK fighter pilot who liked throwing the heavy Wapiti about, which came in useful here because our bombing approach had to be made almost down the side of the nearest mountain. The way he flew towards that mountain, then flipped over in a tight bank to go down onto target scared the Hell out of me! I was convinced we were going to hit the side of that mountain, and I still swear our wingtip only missed doing so by less than a foot!*

With the threat of air and land action against any tribe known

* The late R.A. Wilkins, correspondence with author, 1968.

Wapiti K1291 of 27 Squadron, 1936, the aircraft flown by succeeding squadron commanders, and bearing the Elephant unit badge on its fin.

Miranshah Fort, viewed in 1933.

Wapitis of 60 Squadron at Miranshah during 1938 operations. In foreground is the underground bomb store

to be harbouring the Fakir of Ipi, he was of necessity to change his 'residence' fairly frequently in late 1937 and early 1938, but he continued his propaganda unabated among remaining sympathisers, these including tribes in the southern province of Afghanistan who provided him with stocks of ammunition and other material help. By August 1937 the Fakir was situated in main Mahsud territory south of Razmak, then in the same month moved to the heavily wooded and mountainous area of Shawal in north-western Waziristan where he was joined by some 6,000 tribesmen loyal to his cause, mainly from the Mahsud and Wazirs.

Here his activities were supported by a *mullah*, Sher Ali, from September who led a *lashkar* in attacking a Scout post at Tiarza in south Waziristan, apart from masterminding a series of raiding gangs. Sher Ali's force was duly bombed and temporarily dispersed, but further unrest was instigated by the Bhittanis at the instigation of the Fakir of Din, including the kidnapping of four young Hindu virgins. Again, air operations were carried out on their territory, resulting eventually in their submission to governmental terms (and restitution of the Hindu girls). By October further air action against the Fakir of Ipi's Shawal 'headquarters' forced him to flee to Maddha Khel country further north, after which, with the onset of winter, conditions generally became more settled for a brief period.

The difficulties for all air crews in locating and accurately bombing any proscribed target were not improved by the prevailing meteorological conditions ever-present along the north-west frontier regions of India.

To quote one description:

> Dust haze, though less prevalent than fine weather on the Frontier, nevertheless comes in spring and summer when operations are usually most intense. Thus the familiar sights of innumerable small groups of white-topped tents, picquet signs, scrub, rock, deep-cleft *nullahs* (valleys), lorries winding their way in irregularly spaced batches along the lines of communication, patches of cultivation, and a few small but many-towered villages, are best remembered as seen through

> indifferent visibility with an occasional glimpse of the mountain-crests beyond.*

By December 1937 the Frontier appeared relatively peaceful enough for the RAF squadrons to revert to normal routine – and, of course, to prepare to enjoy the imminent Christmas festivities; always a high point of off-duty pleasure to all Servicemen stationed overseas. The same month also brought the announcement of several awards for RAF personnel engaged in the recent Waziristan operations, including DSOs to Wing Commanders A. Lees† and J.C. Slessor‡, a DFC to Squadron Leader R.H. Carter, OC 60 Squadron,** and the rarer award of a DFM to 27 Squadron's senior sergeant pilot, S.G. Betty.††

A measure of this 'relaxed' period in the context of actual operations was the detachment of complete squadrons to Singapore from mid-January 1938 for Combined Exercises with the RAF and army units resident in Singapore; coincidental with the official opening of the Singapore Naval Base on 11 February. Nos 11, 27 and 39 Squadron joined with 60 Squadron for these particular exercises; all units returning to their Frontier stations by late February (all aircraft somewhat overloaded with 'material delights' from Singapore …).

Almost as soon as the squadrons returned, however, they were plunged back into daily operations. With the coldest weather over, and the festival of *Id* and observance of *Ramadan* duly completed, the approach of spring – the proverbial 'silly season' for tribesmen bent on stirring up trouble – brought further eruptions, part-instigated by the Fakir of Ipi still being sheltered by the Maddha Khel.

Bombing raids against Madda Khel targets commenced on 3 March, including Sar Kalai. The raid on this rather difficult

* 'Through the Waziristan Duststorm', Anon; *RAF Quarterly*, October 1939; Gale & Polden.

† Later Air Marshal Sir Alan, KCB,CBE,DSO,AFC.

‡ Later Marshal of the Royal Air Force Sir John, GCB,DSO,MC.

** Later Group Captain, DSO,OBE,DFC.

†† Later Squadron Leader, AFC,DFM.

objective was led by Wing Commander John Oliver, AFC, OC No 1 Wing with Leading Aircraftman, G.J. Fudge from 27 Squadron as his gunner/bomb-aimer. Making a very low bombing run-in, Oliver's Wapiti was greeted with a virtual barrage of groundfire and was riddled with bullet holes as the bombs were released. Arriving back at Miranshah, Oliver completed a perfect 'three-point' landing – then the undercarriage collapsed. Shattered by the rifle fire, it was by then attached to the aircraft by merely one fitting. Fudge was recommended for the award of a DFM, which was eventually gazetted in November 1939.

These (mainly) anti-Madda Khel operations were to continue into April and May 1938 with increasing intensity as unrest spread widely among the tribes, always encouraged by the Fakir of Ipi's continuing propaganda, including heavy tribal attacks on Scout posts and marching columns which inflicted relatively heavy casualties on both sides. Part of the rebel forces by now were Bahlolzai Mahsuds led by *Mullah* Sher Ali, whose depredations led to a programme of selective and heavily destructive bombing of Bahlolzai Mahsud targets aimed at forcing the non-rebelling Mahsud sections to oust Sher Ali; a ploy which was successful by late May.

By then the Fakir of Ipi had gathered a *lashkar* of a thousand men at Kharre with which he inspired an attack on the Datta Khel Scout Post, using tribal cannons to support the riflemen. Accordingly, an army force two-brigade strong moved over the hills between Razmak and the Tochi Valley to eject the Fakir's *lashkar* which had moved to this area. Assisted by close RAF support, the army finally managed to scatter this *lashkar*, despite having now to fight and march at the hottest season of the year.

With the army and the RAF by then operating at their highest intensity since the 1936 initial Fakir-inspired disturbances, a fresh serious problem arose in addition, threatening to set the whole Frontier ablaze. Sayid Muhammed Sadi, a 34-year-old revolutionary of Syrian extraction from Damascus, married to a German wife and with pronounced pro-Nazi leanings (even rumoured to be directly in Nazi pay), and a

Wapiti K1294, 'D' of 'A' Flt, 27 Squadron releasing a pair of 112lb RL bombs in the Splitoi area, February 1938.

Vickers Valentia K4634, '*CITY OF LAHORE*', 31 Squadron at Kohat 1938. The two mountain peaks just visibile at left were known as 'The Old Lady of Kohat' …

remote relative of the Afghan ex-Amir Ammanulla's queen, Souriya, arrived very unobtrusively in Mahsud territory, ingratiated himself with the tribesmen as a theological arbitrator – settling minor feuds and fixing low prices for marriageable girls *et al* – before showing himself in his true colours. Dubbed 'Shami Pir' from his Syrian origin, he addressed a large tribal *jirga* in the chief Mahsud town of Kaniguram on 13 June 1938, and persuaded three-quarters of South Waziristan tribes that it was their solemn religious duty to restore Amir Ammanulla to the Afghan throne by deposing the incumbent Amir Zahir Shah. In effect, he was proposing an armed invasion of Afghanistan from Indian territory, using Pathan tribes, for an armed coup.

His 'invasion' force set forth for the Afghan border on 20 June, but fortuitously on 22 June, during the course of a normal armed reconnaissance, three Audax aircraft of No 20 Squadron, led by Squadron Leader Basil Embry, AFC,* spotted this *lashkar* in open country very near the Frontier pass at Khand Narai and immediately attacked, creating havoc and no little panic among the tribesmen who put up initial resistance but soon dispersed quickly. Shami Pir with a few companions fled towards Waina where the authorities took him into custody, then by stages deported him back to Syria. The swift nullification of this abortive but pregnantly dangerous revolt was greeted with huge relief by both the Afghan and Indian governments whose common interest in maintaining the stability of the Frontier had been closely affected.

Meanwhile the Fakir of Ipi – who, perhaps, surprisingly, refused resolutely to lend any support to Shami Pir's projected invasion of Afghanistan – continued to stir the tribes to revolt, but his former supporters, the Madda Khel, submitted to government terms by the end of June and evicted the Fakir who fled to Kharre, there to gather new forces. From Kharre the Fakir supervised the replenishment with food, arms and ammunition of a number of tribal gangs who were then

* Later Air Chief Marshal Sir Basil, GCB,KBE,DSO,DFC,AFC who received his first award of the DSO on 13 September 1938, chiefly for his action this day; the first of an eventual quadruple DSO.

dispatched to various points in Waziristan to continue harassing army posts and troops; the centre of such activities being at the Ahmedzai Salient north of Bannu where, on 23 July, a daring raid resulted in arson, murder and heavy looting in the native sections, led by the notorious outlaws Gagu and Mehr Dil, presenting potential danger to the government-administered districts of Kohat and Bannu. Air and ground operations over the Ahmedzai region continued through August and September before the threat was finally nullified.

During these operations No 27 Squadron suffered its last 'peacetime' fatal casualties when on 29 August Pilot Officer G.P. French crashed during landing approach at Mirozai, near Kohat. The Wapiti's petrol tank ruptured, then exploded, killing both French and his passenger AC1 C.P. Rushworth.

Before this latest bout of operations was diminished, events far away in Europe had intruded upon the Indian scene as Europe moved seemingly inexorably towards a war with Nazi Germany as Hitler's demands for yet more *Lebensraum* ('living room') for the German subjects now included Czechoslovakia. The prospect of such a war led to a determined effort by the Indian government to put an end to the Fakir of Ipi's continuing revolutionary efforts, even including an offer of a free pardon for past sins in September 1938, which the Fakir adamantly refused.

The Fakir was to continue to spread and advocate rebellion for years to come, supported yet again in the main by the Madda Khel who reneged on former promises to the Indian government. However, his influence among the tribes was to diminish throughout World War Two and his eventual death was recorded, with characteristic British aplomb, in *The Times* of 20 April that year, despite a description of the Fakir by a leading Pakistan official as 'a vicious old man twisted with hate and selfishness, prone to treachery, and one who gave bonusses for the kidnapping or killing of the children of Political Agents ... '

The Munich Crisis of September 1938 hardly affected the RAF in India apart from a somewhat sudden order for all units to come to a 'readiness' state, with, in the case of 60 Squadron, orders to prepare for a complete move south to Singapore should the Munich conference fail to prevent war. The crisis did, however, serve to highlight the obsolescence of nearly all squadrons' aircraft equipment. In early 1938 every squadron was still operating biplanes of near-vintage designs, the most ancient of these being the Wapitis which were still being flown by Nos 5, 11, 27, 31 and 60 Squadrons. No 39 Squadron, one of the two first Wapiti units to serve in India, had exchanged its aircraft to Hawker Harts some years previously, while Nos 20 and 28 Squadrons had converted to Hawker Audaxes from December 1935 and June 1936 respectively, and the BT Flight was still flying the lumbering workhorse Vickers Valentias.

The trusty Service 'grapevine' was already rumouring the imminent arrival of modern, all-metal, monoplane bombers – Bristol Blenheim Is – and in anticipation of such an event training courses were established at Ambala (air crews) and Karachi (ground crews) for familiarisation on Blenheim aircraft in May-June 1938, while in June 27 Squadron provided two officers and six airmen to form the first Blenheim Collection Party (BCP) which proceeded to Egypt in late June in order to ferry Blenheims back to India.

This first BCP eventually joined 27 Squadron on 26 July, having participated in the first-ever flight by Blenheim aircraft over the North-West Frontier Province of India. The first squadron to convert to Blenheim Is in India was No 11, beginning in July 1938, but it was not until 1939 that the next two units began conversion to the type; No 60 Squadron was slowly re-equipped from March to September that year, followed by 39 Squadron from June 1939. The five remaining squadrons were left to soldier on with outdated Wapitis, Audaxes, and Valentias until 1940-41 before being re-equipped with more modern aircraft.

Despite the continuance of air operations against the recalcitrant tribes along the Frontier, particularly the Madda Khel and Tori Khel, during 1939-40, the RAF in India was inevitably affected to a degree by the gathering storm in Europe where the British government, among others, prepared urgently for the distinct possibility of imminent

OLD & NEW. Wapitis of 27 Squadron with a 60 Squadron Bristol Blenheim I, L1545 at Madras, 1939.

hostilities with Hitler's Nazi Third Reich. The foremost consideration for the UK government in the context of the Far East was the defence of the naval base at Singapore, resulting in the transfer of some resident RAF units on the NWFP during 1939-40.

The first such unit to quit the Frontier area was No 60 Squadron which left Kohat on 2 March 1939 and flew to Ambala, via Lahore – the finale to 19 years of seldom-ceasing operations for the unit along India's northern borders. It was followed in June 1939 by No 39 Squadron which set out from Risalpur on 6 August 1939 with nine Blenheim Is on the first leg of its intended transfer to Kallang, Singapore. Led by Wing Commander Burton Ankers, DSO,DCM, this Blenheim flight proved disastrous, taking thirteen days for the overall journey and losing six aircraft and three crew men killed, including Ankers, en route due mainly to atrocious weather conditions encountered. The squadrons still remaining in northern India found the pace of anti-tribal operations dwindling steadily during 1939-41, and these units were eventually re-located in various war theatres as war progressed.

The coming of World War Two to India and the Far East was, in a sense, the end of an era in RAF history, inasmuch as the years 1918-39 had been the ultimate decades of British imperial responsibility along the North-West Frontier Province. India now became one huge theatre of war, under joint Allied military direction, while very soon after the defeat of Japan India finally achieved full independence as a nation, and the NWFP became merely part of a newly-created Pakistan. When India became legally independent, as from 14 August 1947, and the long-standing British presence began to be evacuated, the tribesmen in the northern provinces were frankly bewildered, unable to fully understand why their traditional fighting opponent, the British Army, was now relinquishing territories it had occupied, bloodily defended, and never been decisively defeated in for more than a hundred years. Moreover, it had successfully thwarted all attempts by Afghans, Persians, Russians and other aspirants to invade and plunder Indian soil. Fierce in battle, offering no quarter to any foe, the Pathan nevertheless held a worthy foe in high esteem, and none higher than the British 'Tommy'. Now, without the British to fight, the Pathan's life would 'lose its savour' *(sic)*.

For the men of the RAF who had served on the Frontier during the years 1918-39 – most of them enduring five years or more there – it had been an unforgettable part of their lives, never to be repeated. A period of real hardships, deprivation, separation from kith and kin for a majority, yet also years of true comradeship and united purpose which would never be surpassed and rarely equalled in any other facets of Service or civilian life. Service on the Frontier had brought most men into direct contact with the deaths of close friends and acquaintances, often in horrifying circumstances, yet such tragedies had served curiously in binding even tighter the communal spirit of all units.

CHAPTER ELEVEN

Sideshows

While the prime theme of this book has been selected examples of actual warlike operational employment of the RAF overseas during the so-termed peace years between the two major global wars of this twentieth century, the description 'operational' in any RAF context can readily be applied to myriad other RAF commitments during those adolescent years of the Third Service. Many such undertakings followed Trenchard's view that his overseas squadrons were, where at all possible, to be used to *prevent* uprisings rather than merely punish dissidents. As he expressed it in a lecture to a gathering of Cranwell cadets on one occasion: 'When you bomb a man's home and family, you can hardly expect him to proffer a hand of friendship.'

Ever conscious of the necessity to present 'his' air force as more than a weapon of pure destruction – the major charge brought by the Service's many political enemies during the 1920s particularly – Trenchard also strongly encouraged his squadrons to 'expand horizons' in various ways. Thus, throughout the 1920s and 1930s normal firstline squadrons and, in certain cases, specially formed RAF units, undertook 'long distance flights' *(sic)* as almost routine training exercises, though with the serious purpose of establishing and/or extending air communications and routes which in turn provided valuable foundations for British civil aviation's global potential.

Allied to Trenchard's contention that the RAF could substitute for cumbrous, expensive army garrisons in controlling overseas areas was the efficacy of the RAF in providing rapid reinforcement and reaction to any form of imminent crisis; the Chanak Crisis providing an early example of the RAF's 'instant mobility' (*sic*). A further occasion in this context occurred in early 1927 when the city of Shanghai came under threat of invasion and occupation by Cantonese Chinese revolutionaries of the Kuomintang movement, led by Dr Sun Yat Sen, when in January these communist-Chinese forces over-ran and sacked Hankow, some 400 miles up the Yangtse river from Shanghai, then created outrages in Nanking and were advancing to Shanghai by March.

A tri-Services' force, titled 'Shanghai Defence Force' was hurriedly despatched from the UK in January 1927 to protect British lives and property in the Shanghai Treaty Port Area. The air sections of this force were initially Fleet Air Arm Flights aboard the China-based HMS *Hermes*, and ex-UK seaplane carriers *Argus, Vindictive, Enterprise* and *Tamar*, but at Shanghai there were no facilities of any description for operating aircraft.

Permission was obtained to reconnoitre a possible landing strip on the nearby Kowloon Peninsula – the 'birth' of the present Kai Tak aerodrome – but within Shanghai the only feasible site was the Recreation Ground in the centre of the Settlement which was first occupied by No 441 Flight, FAA on 14 April when six Fairey IIID aircraft, commanded by Captain G.E. Wildman-Lushington, RM (later Major-General) were stationed there. The only safe landing run was some 400 yards long, parallel to the race-course grandstands, while landing approach was over the crowded city of Chapei, *between* two ten-storey buildings, then over the main Shanghai thoroughfare. Aircraft accommodation consisted of bamboo hangars built by local Chinese contractors, some 80 feet by 70 feet, each costing about £120.

The air elements of the Shanghai Defence Force, now collectively titled 'Royal Air Force China' and commanded by Group Captain E.D.M. Roberston, DFC, received a further addition to its strength, No 2 Squadron RAF, based at Manston, Kent, commanded by Squadron Leader W. Sowrey, DFC,AFC (later Air Commodore, CBE,DFC,AFC) and equipped with Bristol F2Bs. No 2 Squadron had been quickly mobilised and its personnel shipped from Southampton aboard SS *Neuralia* on 20 April 1927, arriving in Shanghai on 30 May and disembarking two days later. With rebel Chinese troops all round the Settlement, and a distinct possibility of mass mob violence from within Shanghai's Chinese population of over two millions, the RAF complement – the equivalent now of about five normal squadrons in aircraft with some 700 personnel – attempted to maintain regular air patrols around the Settlement. This inevitably involved flying over Chinese territory, resulting in loud protests from the Chinese regarding territorial rights *et al*, but provoked no direct actions. One Bristol F2B of 2 Squadron forcelanded on a Chinese race-course outside the Settlement boundary due to engine failure but after various threats and counter-threats from the respective authorities the crew, the F2B's engine and instruments were returned safely, although part of the airframe was retained by the Chinese – a matter of 'saving face'.

Eventually Nationalist Chinese forces assumed control of the Shanghai area and the need for the Shanghai Defence Force, including 'RAF China' diminished. Accordingly, HMS *Argus* and No 2 Squadron returned to England, with 2 Squadron taking up residence at Manston again by the close of October 1927. In the aftermath a small party of men, technical and administrative, were left behind at Kai Tak tasked with preparing the site as a permanent airfield for future possible use. Although RAF aircraft had only been used to provide air reconnaissance roles for co-operation with the army – which had no means of knowing what was happening outside the Settlement – the mere presence of the aircraft had materially prevented a repetition in Shanghai of the outrages in Nanking and Hankow, and probably avoided thereby any form of all-out operations or even war.

Perhaps the biggest example of an RAF overseas reinforcement came in late 1935. The Italian dictator Benito Mussolini had decided that 1935 was to be the 'year of destiny' (*sic*) in East Africa and built air bases in the colonies of Eritrea and Somaliland, with vast supplies of fuel and munitions. Quickly, Abyssinia became surrounded by a chain of airfields, some fitted with concrete runways and permanent hangars as at Gura and Asmara obviously intended as bomber bases.

Then, despite nine months of debate and attempted dissuasion at the League of Nations, Italy invaded Abyssinia in October 1935, and the only response from the League was to call for sanctions against Italy (although it did not include in these vital oil supplies to feed Mussolini's war machinery ...).

Britain, alarmed at the potential threat to its interests in British Somaliland, Sudan, Kenya, Aden and nearby territories, apart from the further interests in and around the Mediterranean, quickly arranged for army, navy and RAF reinforcements for its various Middle East garrisons. In the case of the RAF such additional strength amounted to a total of eleven UK squadrons, apart from diversion of existing units in the Middle East to unfamiliar bases. These squadrons were:

Sqn	*UK base*	*M/East base*	*Aircraft*
3	Kenley	Khartoum	Bristol Bulldog II
12	Andover	Khormaksar	Hawker Hart
29	North Weald	Amriya	Hawker Demon
33	Upper Heyford	Mersa Matruh	Hawker Hart
35	Bircham Newton	Ed Damer, Sudan	Fairey Gordon
41	Northolt	Khormaksar	Hawker Demon
74	–	Hal Far, Malta	Hawker Demon
142	Andover	Aboukir	Hawker Hart
204	Mountbatten	Aboukir	Supermarine Scapa
207	Bircham Newton	Ed Damer, Sudan	Fairey Gordon
210	Pembroke Dock	Alexandria	Short Singapore

All these moves took place in September-October 1935 in a shroud of secrecy – the British public were totally ignorant of this sudden depletion of its metropolitan air defences until the following year – and none more so than No 74 Squadron. Originally disbanded in England in July 1919, the squadron remained disbanded until September 1935 when on 3 September* a collection of ten officers, sixteen NCOs and 75 airmen withdrawn from various RAF stations and units in the

* A retrospective date because the title 74 Squadron was not *officially* allotted until 14 November 1935; the nucleus was known as the 'Demon Flights' until then.

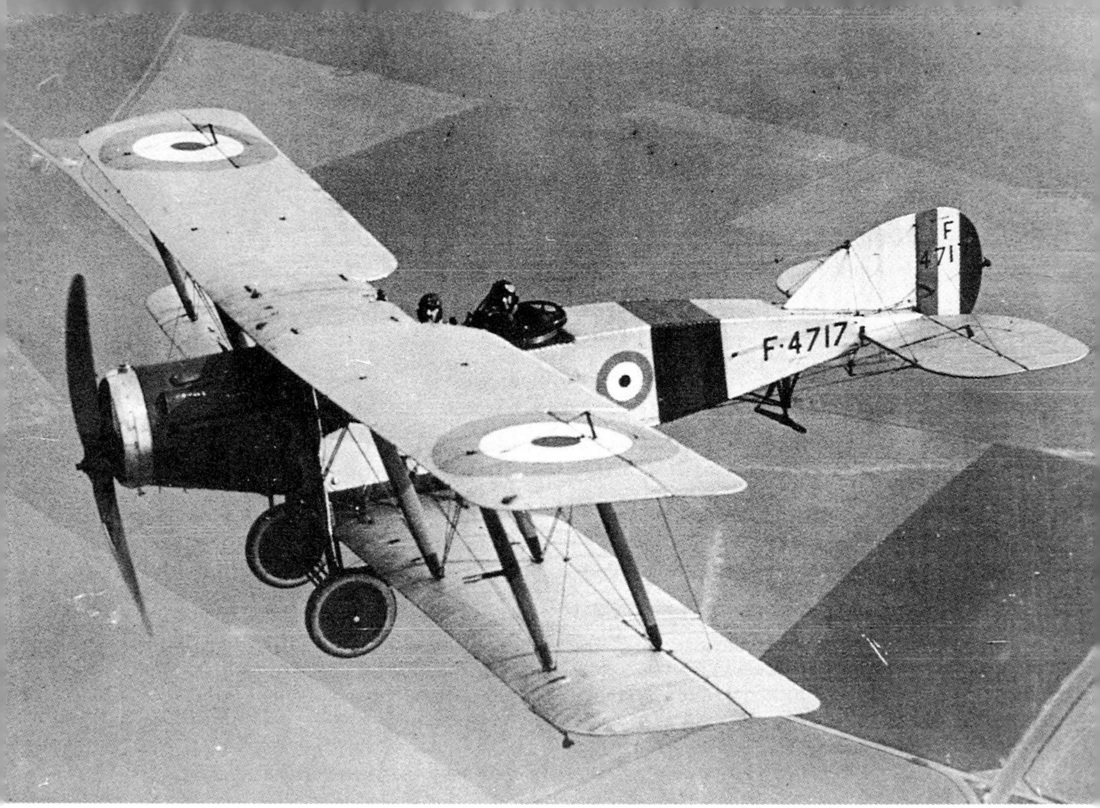

Bristol F2B, F4717 of No.2 Squadron, based at Manston, Kent.

Camouflaged Hawker Demon of No.74 ('Tiger') Squadron on Malta during the Abyssinian Crisis detachment, 1935-36.

UK and embarked in the troopship *Neuralia* at Southampton became the reformed 74 ('*Tiger*') Squadron. None of these, except the appointed CO, Squadron Leader H.G. Crowe, MC (later Air Commodore, CBE,MC) knew their destination, Malta, while the unit's intended aircraft, Hawker Demons from several other squadrons still retaining their original unit markings, were crated and loaded aboard SS *Maihar*. Personnel and aircraft finally came together at Hal Far aerodrome by mid-September, and were operationally ready by the end of the month.

In addition to the input of ex-UK squadrons, No 203 Squadron, normally based at Basra, had just re-equipped with Short Singapore flying boats and was diverted to Aden for the crisis period, while existing units in Palestine, Egypt, Aden and the Sudan were strengthened by the addition of a fourth Flight apiece. The UK squadrons – most had left a few personnel behind as nominal 'base staff' – were to remain on the *qui vive* until August-September 1936 before being returned to England, during which year all had accumulated many useful hours of 'tropical experience' in training and, in certain cases, actual anti-tribal operations alongside the resident squadrons.

The Abyssinian Crisis was not the only problem for Britain (and other nations) during this time, there being ramifications of the Spanish civil war with which to contend, particularly in the areas nearest Gibraltar. Several incidents of Spanish guns, land and shipborne, firing at RAF aircraft and British shipping in the Straits occurred, while British and French patrols by sea and air were instituted to counter piracy by Italian submarines off Spanish ports. In the context of air strength, the Mediterranean itself was chiefly the concern of the Royal Navy and the Fleet Air Arm, with just No 202 Squadron with its flying boats (Scapas) based at Kalafrana, Malta in 1935-36, although No 210 Squadron became temporarily based at Gibraltar before eventually returning to the UK in August 1936.

Such emergencies cropped up from time to time throughout the years 1918-38 in various locations around or near British-controlled territories – Egypt, Cyprus, Somaliland,

Officers of 74 Squadron, Malta, 1935, with the squadron commander Sqn Ldr H G Crowe 5th from left at Hal Far.

The temporary airfield at Gebeit, Sudan, base for an RAF Wing commanded by Grp Capt Ray Collishaw, 1935-36, during the Abyssinian emergency.

Kuwait come to mind readily as centres of unrest and potential revolts – and local army and navy forces increasingly depended upon the RAF to transport troops and supplies rapidly to the troublespots; a task usually undertaken by the few RAF transport squadrons available, notably 70 and 216 Squadrons in the Middle East and India.

Mention has been made in a previous chapter of the well-publicised air evacuation of the Kabul legations in Afghanistan in 1928, but this had not been a solitary incident. RAF aircraft had flown similar emergency evacuations of British residents from various places before the Kabul airlift, and were to do so on many occasions thereafter; more examples of the flexibility and efficacy of air power, albeit on a small scale in most cases.

Equally, the RAF transports were directly responsible for saving the lives of unnumbered operational casualties of the army, RAF and even warring tribesmen over those years; mercy missions able to convey seriously sick people from action areas to backwater hospitals in mere hours, where transportation by traditional land routes might have taken days or even weeks. All these, and other, rarely publicised activities were accepted by RAF crews as simply routine duties without demur; although the RAF's official motto was *Per Ardua Ad Astra* – and probably never more so than in the 1920s and 1930s – it might equally have been *Ubique* ...

Apart from actual operations of a warlike nature, the most significant form of RAF activities during 1918-38 were the many so-termed 'long distance flights' undertaken by most squadrons, both in the UK and overseas-based units. Each in its way was truly a pioneering venture, albeit as carefully planned as contemporary knowledge permitted. Regarded officially as training flights, these aerial probes into what in most cases was the unknown progressively extended the RAF's horizons in literal fashion, but at the same time blazed aerial highways which would eventually become foundations for regular civil air routes around the British Empire. The RAF's immediate purpose for such flights was communication between important outposts and cities in each country concerned, with the added bonus of establishing mobile capability for moving complete squadrons quickly from one base to another should such a need arise. A further reason was the constant task of aerial photography to assist in accurate mapping of lands which, in the 1920s and '30s, had often not seen human occupation within recorded history – uncharted deserts and jungles still the domain of raw nature untrammelled by *homo sapiens*. Such flights called for a degree of courage and self-dependence from the young air crews usually associated with men of more mature years, yet they never failed to accept the challenge, or the potential risks concomitant with any form of pioneer expedition.

These 'long distance flights' varied enormously in scope and intention, and indeed in pure distance through the years, being inevitably confined to the limitations of the types of aircraft involved, availability of staging points where refuelling might be possible, and the type of country or seas to be overflown. With the gradual improvement in aircraft designs through the years such flights could be undertaken to yet more ambitious lengths, including several individual types of machine specifically created for this type of operation. In the main though, a majority of these flight epics were flown by standard RAF aircraft, slightly enhanced to cope with such matters as extra fuel tankage, and crewed by air and ground personnel on normal squadron strength.

To list and describe, even briefly, all such flights, be they titled 'cruises', 'reinforcement', 'long distance', or whatever, would need a separate thick volume. Thus the following selected examples must stand as illustrations of the problems, planning, fortitude and daring involved in all such flights.

In the 1920s, against a background of loudly publicised aerial record flights by various civil pilots, men and women, the RAF quietly set about creating its own schedule of pioneering flights, aimed mainly at improving RAF mobility and providing severe tests for its crews and aircraft. Thus, while most civil air record flights were being accomplished by solo aircraft and pilots, the RAF preferred to undertake formation projects involving three or more aircraft.

An example was the occasion when three DH9As from No 47 Squadron, led by Squadron Leader A. Coningham, (later Air Marshal Sir Arthur), left Helwan on 27 October 1925 and flew to Kaduna, Nigeria, arriving there on 6 November, then returned to Helwan on 19 November, having flown a total mileage of 5,268.

An even more ambitious formation flight was undertaken in 1926. For an especial attempt to establish an air route from Cairo to Cape Town, South Africa – a route already flown by certain individual pilots – four Fairey IIID aircraft (S1102, S1103, S1104 and S1105) and their crews were assembled at Northolt on 6 November 1925 where preliminary consumption tests and practice flights were carried out prior to the Flight being shipped to Aboukir, Egypt. In command was Wing Commander C.W.H. Pulford, OBE,AFC, with pilots Flight Lieutenant P.H. Mackworth, DFC, Flight Lieutenant E.J.L. Hope, AFC and Flying Officer W.L. Payne, while the techincal staff were Flight Lieutenant L.E.M. Gillman, Flying Officer A.A. Jones, and Sergeants Gardener and Hartley; Gillman being a navigator.

An extra Fairey IIID was sent to Aboukir as reserve, while two spare engines and various components were sent ahead to Kisumu, and a second aircraft, four engines and various spare parts were divided between Cape Town and Pretoria. Assembled at Aboukir, the four principal IIIDs moved to Heliopolis, complete with 'overloads' weighing roughly 450lb in each aircraft (spares, tools, medical kits, arms, rations etc). Then at 0710 hrs on 1 March 1926 the four IIIDs set off from Heliopolis, adopted formation, and flew to Cape Town, following the route Cairo-Assiut-Aswan-Wadi-Halfa-Atbara-Khartoum-Kosti-Malakal-Mongalla-Nimule-Kisumu-Mwanza-Tabora-Abercorn-N'Dola-Broken Hill-Livingstone-Bulawayo-Pretoria-Johannesburg-Bloemfontein-Beaufort West and Cape Town. Two propellers and a magneto needed to be changed en route, while at Kisumu the aircraft received a normal 25-hours' overhaul. From Pretoria to Johannesburg the Flight was given an honour escort of 14 aircraft from the South African Air Force, and Cape Town was finally reached on 12 April.

One minor emergency measure during the Abyssinian Crisis was the modification of a Vickers Valentia of 216 Squadron to incorporate air gunners' cockpits in the nose (here) and fuselage mid-ship locations – innovations constructed locally.

'A LITTLE PIECE OF MHUTTI …'. Robat, Aden, the sparse 'home' for No.12 Squadron during the Abyssinian emergency.

Supermarine Scapa, K4192 o 202 Squadron, seen here on ' May 1935 on arrival at Malta (the first Scapa to reach the unit), flown in by Flt Lt C W Dicken and Sgt Otter.

Here four specially built 20-gauge copper tanks were fitted for the return journey which began on 19 April when the four Faireys flew to Pretoria, arriving on 22 April and remaining there until 3 May while aircraft were overhauled and a fresh timetable planned to make the rest of the return flight via Nairobi. On 3 May the foursome took off from Pretoria and eventually arrived back at Heliopolis on 27 May (a day ahead of the planned timing). Total flying time for the four aircraft for the round trip to the Cape and back had amounted to some 570 hours, at an average air speed of 79 mph. The trip back to Cairo encountered bad weather, including sandstorms, on some parts of the course, while a snake which coiled itself round the spokes of the leader's port wheel at Nairobi was flown to Kisumu without apparent harm!

Back in Egypt the four Fairey IIIDs had their wheeled undercarriages changed for float empennages and thus fitted continued their flight by easy stages back to England, arriving at Lee-on-Solent on 21 June. Here they had completed between 1 March and 21 June a total of some 14,000 miles in about 180 actual flying hours. These Cairo-Cape and Cairo-Nigeria flights were soon to become annual or biennial Service exercise flights respectively for Middle East based RAF squadrons.

Undoubtedly the most outstanding long distance and endurance flight of the 1920s for the RAF was the first formation flight from England to Singapore in 1927. In July 1926 two Supermarine Southampton flying boats had flown a 7,000-mile round trip 'cruise' from Plymouth to Egypt, then Cyprus, and back to England in less than 30 days, and this extended 'test' of the Southampton aircraft led to the type being chosen to equip an even greater project, to fly to Singapore. Four such aircraft were selected for this venture, specially built and modified, while a fifth Southampton S1127 was to be sent on to Singapore by ship to act as a spare if required. Chosen to command this Far East Flight, as the project was titled, was Group Captain H.M. Cave-Browne-Cave, DSO,DFC (later Air Vice Marshal, CBE,DSO,DFC) and the other three flying boat captains were Squadron Leader G.E. Livock, DFC (later Group Captain, DFC,AFC), Flight Lieutenant

D.V. Carnegie (later Air Vice Marshal CB,CBE,AFC) and Flight Lieutenant C.G. Wigglesworth, AFC.

By May 1927 all personnel had been gathered together at RAF Felixstowe; four flying boat crews consisting of two officers and two airmen in each, plus a base party comprised of three technical officers and 23 airmen. Their brand-new Southamptons, S1149 (Livock), S1150 (Carnegie), S1151 (Wigglesworth) and S1151 (Cave-Browne-Cave) were delivered in good time and were metal- hulled Mk IIs fitted with Napier Lion engines of 450hp each. Cockpits were standard open-type, no automatic pilots or blind- flying instrumentation were fitted, and no armament carried apart from Very signal pistols. Each Southampton was intended to be self-dependent in terms of crew accommodation, cooking and sleeping arrangements, toilets, personal kits, tools, etc.

The overall objectives of the flight were partly 'showing the flag' to all countries en route, collecting information on possible seaplane bases and harbours and local conditions, and testing the ability of a detached Flight of flying boats to be independent of vessels and shore bases, exercising their own initiative and resource under widely varying conditions.

On 14 October 1927 the four Southamptons flew from Felixstowe to Plymouth (Mountbatten) from where at 0900 hrs on 17 October they flew off Plymouth Sound on the first leg of their global journey, arriving at Marseilles on 19 October. The route to Singapore was flown via Naples-Aboukir-Baghdad-Karachi-Bombay-Cairo-Madras-Calcutta-Rangoon-Singapore, and on 28 February 1928 all four Southamptons glided down for a formation landing at RAF Seletar in the Johore Straits – completing the first RAF flight from England to Singapore. This was only the first stage however. The Far East Flight next prepared to fly on to Australia where it would literally fly right round the Australian coast-line and return to Singapore.

This 'second stage' was scheduled to start in May, and meanwhile the Southamptons were given a thorough overhaul at Seletar which at that period had no permanent buildings, no hangars, and personnel accommodation being of native attap construction was decidedly primitive, leaving the flying boats and their crews literally in the open suffering the effects of a

Fairey IIIDs of the first Cairo-Cape flight, on return to Heliopolis, 1926.

Flt Lt S D Macdonald, 47 Squadron testing the first Fairey IIIF, S1141 prior to the 1927 Cairo-Cape flight.

Supermarine Southampton, S1151, of the Far East Flight prior to the epic venture.

tropical sun as work on maintenance proceeded. The journey to Singapore had covered some 9,500 miles in 140 hours of actual flying, yet the aircraft were found to need little refurbishment; indeed even the Napier engines needed no change for the Australian trip.

On 21 May the Flight set out for Australia and eventually touched down at Broome on 1 June. Circumventing the Australian continent was completed without mishap and on 1 September the Flight left Darwin to return to Singapore, reaching Seletar again on 15 September. The crews had now spent eleven months on their 'cruise' and had already covered some 20,000 nautical miles, but there was yet a third 'stage' to complete, a flight to Hong Kong and back to Singapore.

For this last trip Southampton S1127 was substituted for S1149 and the Flight left Seletar on 1 November, cruising via Borneo and the Phillipines to Hong Kong where it remained from 18-29 November, then flew down the east coast of French Indo-China to Saigon, up the Gulf of Siam to Bangkok, across the Malay Peninsula to Victoria Point and down the west coast of Siam and Malaya back to Seletar where the cruise ended on 11 December 1928. The cruise overall had lasted fourteen months and covered 27,000 miles – about 3½ times the circumference of Earth at the equator – without any serious incidents to either aircraft or personnel; a triumph of planning and airmanship hailed in the contemporary press as 'The greatest flight in history'.

The Far East Flight then remained at Singapore and on 8 January 1929 was retitled No 205 Squadron RAF, commanded initially – appropriately – by Group Captain Cave-Browne-Cave, who was succeeded in command a year later by his ex-deputy on the Far East Flight, Livock.

Ten years after the Far East Flight's epic cruise, Australia was to become the venue for yet another outstanding feat of endurance. On 1 January 1938 a special formation was formed at Upper Heyford to explore the limits of Service aircraft ranges, with the eventual aim of capturing the world record (from the Russians) of long distance, nonstop flight in a 'straight line' (*sic*). Titled RAF Long Range Development Unit and commanded by a long-distance veteran pilot Wing

Commander O.R. Gayford, DFC,AFC, the LRDU was issued with specially modified Vickers Wellesley bombers, four of which left Cranwell on 7 July 1938 and flew to Ismailia, Egypt, covering the 4,300 miles in 32 hours flying time (the longest nonstop formation flight to date).

Ismailia was selected as the most favourable start point for the attempt on the Russians' world record of 6,306 miles, set the previous year, and on 5 November 1938 at 0355 hrs three Wellesleys set off for Australia. These Wellesleys were L2638 (piloted by Squadron Leader R. Kellett, leader)*, L2639 (Flight Lieutenant H.A.V. Hogan)† and L2680 (Flight Lieutenant A.N. Combe)‡. Each Wellesley carried some four tons of petrol – roughly half the overall take-off weight – and climbed to 10,000 feet, the planned cruising altitude which was maintained throughout the flight. Below them, stretching all along the intended flight path were seaborne and land signal staging points to aid navigation.

On the second day, traversing the Bay of Bengal, the trio ran into atrocious weather conditions of headwinds and severe thunderstorms. These began breaking the fabric skins of wings and fuselages, while at one point Combe's aircraft became 'lit up' (*sic*) by static electricity with St Elmo's fire playing around it in spectacular display. All three aircraft survived these conditions with little damage but on reaching the Dutch East Indies it was realised that Hogan's Wellesley might not have sufficient fuel left to cross the Timor Sea and reach Darwin with a reasonable margin of safety.

Accordingly, Hogan landed at Koepang (Kupang) to top up fuel, while the other two Wellesleys continued to Darwin, landing at Ross Smith airfield there at 0400 hrs GMT on 7 November, having completed a flight of 7158.7 miles – a new world record – in slightly more than 48 hours; a record figure which was to stand for the next eight years. Hogan followed slightly later and rejoined the unit at Darwin.

An indication of the precise calculations involved in the two Wellesleys which completed the flight first was the remaining

* Later Air Commodore, CBE,DFC,AFC
† Later Air Vice-Marshal, CB,AFC
‡ Later Group Captain, AFC

fuel in each aircraft after landing at Darwin; 44 Imperial gallons in L2638 and merely 17 gallons in L2680. On 24 January 1939 their flight was officially acknowledged as a figure of 7158,653 miles (11,520.421 km). This achievement was rewarded by the awards of six Air Force Crosses and one Air Force Medal to the crews, including one AFC to Flight Lieutenant Heffernan, RAAF who had been the crews' official liaison officer during their stay in Australia. Another honour was the presentation of the Royal Aero Club's Britannia Challenge Trophy for the year's most oustanding example of British airmanship.

Although such flights received well-deserved public and press acclaim they really represented the tip of an iceberg in the context of the many other wide-ranging duties carried out by normal RAF squadrons during 1918-38; a case in point being the numerous occasions when RAF aircraft became air ambulances to evacuate sick and wounded Servicemen and civilians of all nations from remote locations in Middle East lands. By 1939 it had been calculated that some 2,600 patients in all such territories had been evacuated by RAF aircraft in flights covering at least 320,000 miles; details of many such mercy flights being well detailed in the book *But Not in Anger* by C.C.H. Cole/R. Grant, published by Ian Allan Ltd, 1979. Other mundane but vital services undertaken by RAF crews included establishment of regular air mail deliveries in every major theatre of Service operations, aerial surveys of uncharted areas, army troop reinforcement transportations, air supply-dropping of needed supplies to forward army fighting units, and sundry other such examples of air support. Many such duties, particularly during the early years, were completed despite myriad hazards of climate, unreliable aircraft engines, primitive facilities, while in such areas as Iraq and the Indian North-West Frontier a forced landing in wild, rebel-infested country often resulted in death in any of its most horrifying means. All such 'sideshow' operations provided the crews – air and ground – with a hard core of invaluable experience which was to be embellished and expanded when the RAF was called upon to wage a major global war for the second time in its brief existence in 1939.

CHAPTER TWELVE

Kaleidoscope

Any man who served with the Royal Air Force overseas during the years 1918-38 would confirm the view that his daily life was too often a tedium of boredom and frustration, only occasionally enlivened by periods of intense labours and frightening actions. Individual memories of those tours of duty overseas – as in all human nostalgia – usually recall first the 'good' times; close friends, particular celebrations and triumphs, the tight-knit *esprit de corps* of an operational squadron community.

Such warmly remembered facets are, however, inevitably mingled in the mind with the 'other side of the coin'; seeing close pals burnt to death or blown to pieces in fatal flying accidents, or simply 'missing' among the wild mountains and deserts of tribal territories with no subsequent hint of their fates.

Then again there were the myriad humorous incidents – situations, people and events which appealed to an airman's esoteric form of twisted humour – which provided a light balm to otherwise grim conditions. Probably the sharpest memories were of individuals who could rightly be termed 'characters'. In an era when men were raised to be self-dependent rather than relying on a welfare state to provide their living, the RAF abounded with rugged individuals of what might now be regarded as 'eccentric' natures. Such characters were to be found in all levels of rank and trade, each adding his small niche in the annals of RAF lore and legend.

In the period 1918-38 (and for that matter, many years thereafter) any airman or officer's introduction to his first tour of duty overseas was via one of His Majesty's Troopships (HMTs) – air trooping as a norm was not introduced until many years later. Among the best-remembered HMTs were the *Somersetshire* (see Appendix 2), *Dorsetshire, Prinz Ludwig, Brandenburg, Glengorn Castle, Oxfordshire, Assaye,* and others whose titles became notorious in all airmen's memories as epitomising the nadir of discomfort allotted to 'other ranks' (as all non-commissioned airmen were officially titled). Such conditions aboard an HMT are succinctly described by Air Vice-Marshal A.G. Dudgeon, CBE,DFC who was 'trooped' by sea to India in the late 1930s:

> Conditions for the 2,000 troops aboard [HMT *Dorsetshire* in this case] were reminiscent of a mid-eighteenth century prison-ship. Six levels of wooden planks had been fitted in the holds to make decks. These had beams roughly six feet apart. From these beams the soldiers and airmen slung navy-type hammocks each night and took them down again each morning. They were so close together that you could not walk between them – you had to crawl underneath … The officers were luckier and shared a normally two-berth cabin among four.
>
> Also, being amongst pilots, my colleagues were not sea-sick. During the two days and three nights of storms in the Bay of Biscay, the OC Troops had delegated all Orderly Officer's duties to the Air Force. Only they had stomachs so toughened by aerobatics that they could get round the troops' decks without being sick. Even then it was touch and go as troops' vomit with its nauseous stench dribbled through the gaps between the boards on to the hammocks of the decks below. For the men themselves, confined there by the storm and with no effective ventilation, it was horrendous. In good weather, when all troops came on deck, none could sit down for lack of room; all had to stand. For the rest of the day, fresh air was taken in shifts. Officers were allowed enough deck space to get fresh air at any time.*

Trooping by sea was carried out mainly twice per year, in the

* *The Luck of the Devil* by Air Vice Marshal A.G. Dudgeon; Airlife, 1985

so-termed 'cooler' seasons, conveying drafts of men both to and from the various overseas countries, England to India, disembarking men at Egypt, Iraq and Aden en route, and vice-versa for tour-expired men returning to kith and kin in the UK. In between these trooping seasons small parties of men posted overseas were often embarked as civilian passengers aboard normal civil liners and ships, while occasionally in the 1930s officers posted to Iraq were sent by road and rail across Europe. All troopships were organised and run like a military land establishment, with an appointed OC Troops and a plethora of routine daily duties – guards, mess orderlies, deck orderlies, orderly officers, orderly sergeants, *et al* – and Mess Dress for officers at every evening meal.

In addition there were normal ship's routines to be borne – daily lifeboat drills, Captain's Rounds (an eagle-eyed inspection of every nook and cranny), Church Parades each Sunday, and spells of PT (Physical Training) – leaving little time for relaxation. Washing facilities for the 'ORs' (Other Ranks) were basic and minimal and usually in salt water, while toilet facilities were crude. Any innocent airman or officer anticipating the equivalent of a luxury cruise through the sun-lit 'tropics' was rapidly disillusioned within hours of embarking at an English port of departure!

The basic overseas tour of duty for most ranks was five years from early 1924, though this stint was often split by serving two years in Iraq, or three in Egypt, with the remaining time in India, although some posted direct to India spent their complete tours there; the 'choice' of posting being wholly a whim of the RAF Records Office in England. On arrival at the appropriate port of disembarkation, the newly-arrived airman and officer still faced a weary land journey in many cases to finally reach his specific posting. Those posted to a unit on the North-West Frontier Province of India, for example, spent about a month on board a troopship before disembarking at Bombay, from where they were initially accommodated at Deolali, a tented reception camp nearly 50 miles north-east of Bombay; a camp which had once been the site of a hospital for mentally disturbed Servicemen – hence the expression

'Dooolally Tap' used by all Services for those men whose prolonged stays in India had finally unhinged their minds.

From Deolali airmen eventually entrained for the NWFP – a journey which varied between 24 and 48 hours according to which RAF unit was the venue. On arrival at his specific unit an airman was quickly made to feel at home by the residents, as described by R. 'Dicky' Bird, a ranker who had completed two years in Iraq with No 55 Squadron before moving on to India to finish his overseas' tour in October 1932, being posted to No 27 Squadron based at Kohat:

> The outstanding thing of note in my first two days in India was that on our journey to Rawalpindi (from the troopship) the travelling arrangements included railway carriages, whereas in Iraq the government could only afford open iron trucks. By the odour emanating from those trucks I should think they were not usually employed for transporting RAF personnel! The journey from the railhead to Rawalpindi was made by Leyland lorry to Risalpur where we stayed overnight, and then continued on to Kohat, again by Leyland.
>
> When I arrived at Kohat the Christmas arrangement were well in hand and we – the new arrivals – were soon roped in to paying in to the 'Bun Club' which helped to provide funds for all the festivities. One Flight's bungalow in each squadron was converted into a bar and refreshment room, and great rivalry existed between bungalows as to which Flight on the station provided the best 'Set Piece'. The 'Set Piece' in our own was called HMT *Just Imagine* – a mock-up of a trooping ship with all the thirst-quenching drinks hidden in the bridge structure. The rivalry in everyday life between Flights in 27 Squadron was pretty strong, but all banded together if 60 Squadron was involved (also based at Kohat), and when it came to Station against Station, 27, 60 and the HQ combined enthusiastically to see the rival Station off ... It was a period of squadron life for which I hold many happy memories'.*

* Correspondence with author, 1969

'SHIRE, SHIRE ...'. HMT *Somersetshire* of RAF legend passing RAF Calshot in the Solent, bound for India with a full load of Servicemen.

Christmas Day 'Sports' at Abu Sueir, Egypt, 1931.

An airman, H. Howe, who served with the RAF in 1920-26, saw service in Egypt, Iraq and India, and has recorded his blunt views of an airman's life overseas in those early years in a small book, *Over the Way with the RAF* (A.H. Stockwell, u/d). The following extracts give no rose-tinted appraisal of life as an 'OR' then, in his case with No 70 Squadron at Hinaidi, Iraq:

> During the summer months we were allowed to go on parade without a tunic, but during this period we always wore a neck shade that fastened round the *pugarree* (hat-band) of our topee and shaded the nape of our neck, and a spine pad that fastened round the neck and protected the spine as far as the waist line. White canvas shoes, usually acquired from the bazaars of Baghdad, were allowed to be worn, instead of the normal heavy boots, and puttees – detrimental at any time – were discarded. Our topees, the most important and vital part of our dress, were worn at *all* times, and to go outside our huts during the heat of the day without our head-dress on was not only courting death but was a most serious offence, bringing severe punishment to the culprit if caught ... Owing to the terrific heat of the summer in Iraq, men remain throughout the day in a continuous sweat, and without exaggeration it was possible to take your singlet off and wring the perspiration out. Tunics and shorts showed large patches of wet and the least exertion soaked a man from head to foot.
>
> Mespot is a hell country and part of that hell was made by the men themselves, they preferring it to the otherwise cruel monotony of lonely and pious inactivity, alone sufficient to snap a man's mind. One might ask why this excessive drunkenness and gambling was allowed, why NCOs did not stop it. For one thing the country acted on them and the officers as well as men. Another thing, there is 'barrack life', a thing not often tampered with by superiors. NCOs out in this cantonment shut away from the world by a vast desert stretching hundreds of miles, were on many occasions assaulted and terribly manhandled by individual men. It was not an uncommon sight to see NCOs and even sergeant-

majors* appear on parade battered about the face ... It is one of the most serious crimes to strike, threaten or assault an NCO, and anywhere else except in Mespot, if such a thing happened punishment would be swift and severe for the offender, but in this sun-baked desert of Iraq, unmolested from outside authorities, many rules were broken. The peculiar thing concerning such assaults was that the NCOs or perhaps even a sergeant-major never reported their individual cases of assault ...

We were a happy crowd of men who clambered aboard the SS *Glengorn Castle* at Basra (to go home). We were not by any means the same type of men who two years previously had landed there to commence our service in this hell country. Most of us bore traces of hardship, bad living, and illnesses we had contracted during our stay in Mespot. Some of us were like living skeletons and I myself had lost over three stones in weight; a six-foot man and now weighing only eight stone twelve.

No matter what happened off-duty in 'barrack life', however, the prime concern of all airmen – maintaining all squadron aircraft to the highest possible operational standards – was never diminished. All knew that the lives of 'their' air crews were literally in the skilled hands of the Erks, and that responsibility was never skimped in any manner. On the squadrons each aircraft was allotted a rigger and a fitter who, with its pilot, came to be nicknamed the 'Golden Trinity' – a tribute to the close liaison between air and grounds crews, each fully appreciating the other's tasks and efforts. Such close acquaintance never led to over-familiarity or ever corrupted discipline; each *respected* the other, despite the contemporary overt divisions in class.

The constant need for maintenance and refurbishment of increasingly obsolete aircraft to meet operational requirements

* The rank of sergeant-major was retained in the RAF until April 1933 when it was abolished, leaving only the ranks warrant officer Class I and Class II; the latter category of WO II also being eventually discarded by 1940.

often necessitated the 'Erks' working long beyond the officially stipulated working hours, yet such labours were undertaken voluntarily and without resentment, and it was no rarity to find the pilot of a particular machine 'mucking in' with his own ground crew to render 'their' aircraft fit for ops next day.

In India and Iraq particularly, virtually all aircraft were of two-seat (or more) configuration, e.g. the DH9A, Bristol F2B, Westland Wapiti and later Hawker Harts and Audaxes. All required a second air crew man in their rear cockpits, to act as observer-cum-air gunner-cum-bomb-aimer-cum W/T operator. Very few officers were qualified as air observers – wearing the official air crew badge of a winged-O (or 'Flying Arse-Hole' as it was crudely dubbed), thus the bulk of the back-seat men were volunteer ground crew airmen; there being then no official air crew category of air gunner, although some airmen could 'qualify' for the contemporary air gunner's badge, a brass winged bullet badge worn on his tunic sleeve,* which entitled him to a few extra pence pay per day if 'employed on operational flying duties' (*sic*).

Such flying duties were additional to the Erks' normal trade responsibilities and duties, yet there was never a shortage of volunteers. Armourers and wireless tradesmen naturally proved highly adaptable to the job of air gunner, though there was no restriction in trade for those wishing to fly. In the event air gunner, as a distinct air crew category or 'trade', was not officially recognised until an Air Ministry Order (AMO), A17/1939 dated 19 January 1939, followed by a further AMO (A552/39) issued on 12 December 1939, which introduced the air gunner half-wing cloth badge for qualified air gunners, worn on tunics above the tunic's left breast pocket, so familiar on squadrons during the 1939-45.

The late Cyril Anderson was one such 'back-seat-warrior' – an LAC armourer who joined No 27 Squadron on India's North-West Frontier at Kohat in late 1928, just in time to enjoy the 'Grand Christmas Do' (*sic*) organised with talents drawn from both Nos 27 and 60 Squadrons there.

* A badge officially introduced in 1923.

(*Left*) LONG SHORTS OR SHORT LONGS?. RAF Erk in India in the 1930s, complete with topee, spine pad and puttees (*Right*) AIR GUNNER. LAC Wright of 5 Squadron about to fly in a Bristol F2B, Quetta, 1922, wearing a standard Sidcot flying suit.

TOURISTS. Off-duty Erks in an RAF Crossley Tender visit the ancient Arch of Ctesiphon, south of Baghdad, alongside the River Tigris, ca. 1927.

My arrival was greeted with obvious friendliness by the well-tanned airmen who had been on the squadron a long time in India, who helped me to settle in quickly and advised me on such matters as working hours, native servants, dhobi arrangements, and, of course, to beware of 'loose wallahs' (*thieves*). Most of all I was duly warned that the 'Old Man' on 27 *(Commanding Officer*), Squadron Leader Lionel Bailey*, was a 'hard man' who believed in very strict discipline, therefore to 'watch your step' ... 27 Squadron had only just moved to Kohat, from Risalpur, and everyone was very busy sorting things out and getting ready for flying operations again as soon as possible, but the Christmas celebrations went ahead anyway. I soon slotted in to the work and shortly after 27's aircraft [DH9As] were involved in the famous Kabul air lift of hundreds of civilians from over the border in Afghanistan. It was a very busy time for all crews, and everybody mucked in to keep the Ninaks serviceable.

When the hot season was well under way in 1929 I began flying as a gunner with various pilots and was keen to qualify for my Brass Bullet. Being an armourer, handling the Lewis gun on its Scarff Ring in the rear cockpit was no problem, and naturally I knew all about our bombs, but had to learn a bit about W/T work, the Morse Code etc. Our Ninaks were pretty old and well-patched but we had remarkably few accidents, though I can recall several near-misses. On one flight my pilot and I had just completed a routine recce along the frontier and were about to land back at Kohat when at about 200 feet on the final approach the Ninak hit a heat bump. The Ninak reared sharply, its starboard wing lifting fast, and its nose then dropping. I honestly thought my last day had come and gripped the Scarff Ring frozen with fear, but the pilot levelled her out and smoothed in for a normal landing. Climbing out afterwards he looked at me, still gripping the Ring, and laughed. 'Any landing you can walk away from is a good one!'

Actual bombing operations were mostly straightforward

* Squadron Leader (later Wing Commander) L.M. Bailey, AFC, died 2/3/64.

WINGED BULLET. The brass badge for qualified air gunners, seen on this airman's tunic sleeve.

Sergeant S G Betty of 27 Squadron in his Wapiti, 1937 (later, Squadron Leader, AFC,DFM).

affairs; once all the preliminary leaflets and warnings dropped had been ignored, we sent off three or more aircraft to drop bombs on the village in question, by which time the place was empty of the tribesmen and their families. If the target was in fairly open ground, bombing was usually very accurate and we often bombed in formation but we often had to try to winkle out gangs of tribesmen from caves in nearby hills or even on the sides of steep mountains.

For these sort of targets bomb were not always the best solution and the gunners were used to strafe such inaccessible spots. This meant flying very close to the rock-face of the hill or mountain, battling against quite strong up-currents of air, and it was not uncommon for aircraft to return with damaged wing-tips from 'nudging' the rock-face! The more experienced pilots had this 'dicing with death' off to a fine art, but the sprog pilots occasionally scared me out of my tan! I was with 27 Squadron for just over a year before I developed a form of meningitis which in the end saw me invalided home and out of the Service.*

One pilot who joined No 27 Squadron in 1934 as a Flight Lieutenant was Air Vice-Marshal J.M. Cohu, CBE, whose memories of his tour in India were a mixture of unforgettable experiences:

In a crowded memory of the times, first place must be given to a tribute to the NCOs and aircraftmen, the 'other ranks'. Their lot was a very hard one and yet the service they gave was the very best. Another Kipling should write their deeds. Theirs was a tour of duty up to seven years straight, without home leave, without their wives and families except for a small minority, and without female companionship of any sort. In spite of these conditions they gave unstinted service and loyalty at all times. The squadron then was equipped with Wapiti aircraft which were reliable and admirably versatile. In addition to the usual bombing and air firing

* Correspondence with author, 1967

training there was constant reconnaissance over tribal territory and every variety of army/air co-operation. The weather was normally excellent for flying and the principal hazards were the possibility of engine failure and of rifle-fire over tribal territory. There was also the constant danger of collision with the large hawks which pestered the air approaches and caused at least one fatal flying accident in my time. Ransom notes were always carried in case of forced landing to discourage the mutilation and torture in which the tribeswomen had a keen interest. Discipline and high morale combined to foster a standard of professionalism in flying which has probably never been exceeded.

Living conditions were only as good as the times and the situation permitted. The heat in the hot season was a burden to be endured by all. Work started at dawn and finished about one pm when the temperature was near the hundred mark. Sleep at night under mosquito nets was often difficult, and sometimes a hellish wind off the desert sent the temperature as high as 115-degrees F. The women and children fled early in the season to the hill stations where they shivered for a time until the last of the snows melted. The men followed later, each for a few weeks in turn, and as there were so few of the one sex and so many of the other the scandals were frequent – and not always without good cause.

The sister squadron to No 27 at Kohat was No 60, but sisterly regard had no place in the inter-squadron and inter-Flight rivalry which prevailed. There was competition in almost all human activity from flying hours to sporting results. Pilots averaged about 30 hours a month, with even more for the 'hogs'. Every two months a Flight was detached to Miranshah, an outpost on the border of Afghanistan. Here the Flight commander ruled like a captain of a ship and the opportunity was taken to fly hard, as there was little else to do. The season of the year made little difference to the upper air temperature. High level practice bombing height was 14,000 feet and the cold was intense in the unheated open cockpits. However, on the run to the north-east one had at least the privilege of a unique view of part of the highest mountain range in the world. This section of the Karakoram away in Kashmir included Nanga Parbat and Godwin Austin, both well over 25,000 feet. The frontier at this time was relatively peaceful – though the term 'peaceful' could never be truly descriptive of the Frontier Province – and the almost daily sight and sound of our aircraft over tribal territory was usually sufficient deterrent to potential law-breakers. Somewhat rarely (at this period only) live bombing was ordered, to the delight equally of the air crews concerned and of the Station armament officer as a golden opportunity to dispose of time-expired and dangerous to handle bombs. These were usually dropped from 4,000 feet on the mud-walled forts and villages by day, or on any sign of life at night in the proscribed area after ample notice and warning.

Leave could be accumulated for a mid-tour visit to the United Kingdom*, or alternatively, with greater official encouragement, taken in India. With the latter the scope was limitless. For the sportsmen there were all kinds of shooting and fishing, from tiger to small game; for the tourist the Taj Mahal and lesser sights; for the less energetic the pleasures of the houseboat on the lakes outside Srinagar in Kashmir. Kashmir with its many attractions was most popular. There was golf at 6,000 feet at Gulmarg, or trekking in the finest mountain and valley scenery in the world, or the bathing and wining and dining on the Shalimar Bagh. In winter Gulmarg provided super ski-ing which was approached with great numbers of duck on the horizon wheeling like bonfire smoke on the wind, and everywhere there was the aromatic smell of wood fire on the air.

Life on an RAF squadron on the Frontier in the early thirties was for many a time of hard work and hard play, full of interest and new experience. Others found it a hardship with separation from their loved ones and for these the days and months were long indeed. Yet for all it was an

* Officers only – the scheme did not apply to non-commissioned ranks. [Author]

unforgettable experience and a challenge which nearly all rose to meet.*

Cohu's opening tribute in his remarks to the ground crews was equally applicable to the air crews, particularly NCO pilots. In the immediate post-1918 years virtually all RAF pilots were commissioned officers but by 1923 a number of sergeant pilots had begun joining the overseas' squadrons. The first of these to arrive on Nos 27 and 60 Squadrons – three to each unit – in early 1923 were in the nature of an experiment by the Air Ministry, but quickly proved their high value as predecessors to many more excellent senior NCO pilots in the years to come; many eventually rising to relatively senior commissioned ranks and achieving distinguished careers in RAF and civil aviation.

No few sergeant pilots by the 1930s were ex-RAF Aircraft Apprentices, trained as fitters at Halton with several years ingrained experience as ground crews before remustering to flying duties. Part of that experience had been a need for self-dependence, an ability to undertake individual responsibilities in unusual circumstances where 'outside' help was not readily available.

Examples of such self-sufficiency and skills were many among the overseas' squadrons. In Iraq a Wapiti, piloted by a SNCO, had its propeller blade tips badly splintered by loose stones landing at Muscat during a training long distance exercise. The damage was enough (normally) to prevent take-off with safety, but any new propeller for the Wapiti would have entailed it being flown from Baghdad some 1,200 miles away. The sergeant pilot, an ex-fitter, cropped both propeller tips, rebalanced the air screw, flew a brief cautious 'air test', then calmly flew the Wapiti back to base.

A further example of individual SNCO self-sufficiency was a sergeant pilot of No 55 Squadron who had been a Fitter I prior to becoming a pilot. Tasked with an aerial photographic survey of the Iraqi southern desert area, he was detached along to Rutbah Wells, some 250 miles west of Hinaidi, from where he

* Correspondence with author, 1967.

and his photographer/air gunner airman successfully completed surveying several thousands of square miles of the uncharted desert, enabling accurate maps to be drawn for future operational use. These, and so many other examples of individual know-how, epitomised Hugh Trenchard's original vision of 'his' air force applying the high standards he had laid down for training and operations of all facets of RAF service.

The inherent dangers of actual operational flying by the air crews of overseas' squadrons during 1918-38, whether in Iraq, India, Palestine, Egypt or elsewhere, have been described variously in previous chapters. They are also emphasised by the many gallantry awards and decorations received by individual officers and airmen throughout those years, apart from the tragic losses of so many men in flying accidents and other causes.

An early example of gallantry under fire came on 5 May 1921 when Flying Officer V.E. Groom, DFC of No 55 Squadron, based at Mosul, Iraq, took part in a bombing sortie against dissidents and noticed a fellow DH9A shot down to a forced landing in the midst of the rebel tribesmen being attacked. Groom, without any hesitation, quickly landed alongside the stranded Ninak, took its pilot aboard in his own rear cockpit, while the air gunner clambered onto the lower wing of Groom's DH9A and hung on by his fingers. In face of heavy rifle fire from surrounding tribesmen throughout, Groom managed to take off again and bore his 'excess load' back to base. He was duly awarded a Bar to his DFC, and eventually rose to become Air Marshal Sir Victor Groom, KCVO,KBE,CB,DFC.

Other randomly selected examples of such awards for operational flying include one CB, one CBE, two OBEs, two MBEs, three Bars to the DFC and six DFMs for gallantry during just one period of active operations on India's NWFP between 23 April and 12 September, 1930, along with seven BEMs to various ground crew NCOs and airmen. Such awards continued to be promulgated up until the outbreak of World War Two, as in the case of AC1 J.K. Lyon, an RAF wireless operator, who was sent out on 12 June 1939 with an escort of a company of the 2nd Battalion, King's Own Royal Regiment to provide ground liaison with RAF aircraft engaged in anti-Arab operations in Palestine.

As he neared Beit Natif, twenty miles west of Bethlehem, he and the army escort came under heavy fire at some 500 yards range from a rebel gang some 50-strong. Lyon's vehicle, its only 'protection' a canvas cover, remained in the open for two hours under fire, but Lyon remained within receiving and transmitting signals to bring up reinforcements. Lyon was subsequently awarded a Military Medal.

This 'Army' decoration, rarely granted to RAF personnel, was but the latest of a string of similar Military Crosses and Military Medals received by RAF ground personnel throughout the pre-1939 years of 'peace', particularly by men of the RAF Armoured Car Companies in Iraq and Palestine who not only liaised closely with the RAF and the Army on operations, but often undertook active operations independently in quelling local uprisings among the tribes.

The work of the various RAF Armoured Car Companies during 1918-38 has tended to be little mentioned in official accounts of RAF activities then, yet they were a constant factor in operations in the many Middle East zones under British control. The history of the RAF's 'Iron Maidens' went back – officially – to December 1921, when a training unit for armoured car crews was established at Heliopolis in Egypt, which became retitled No 1 Armoured Car Company with effect from 1 February 1922. Equipped with a mixture of Rolls-Royces and Crossley Tenders, No 1 ACC then moved to Iraq where personnel from another armoured car training establishment at Manston, Kent joined the company in September 1922. In April 1922 No 2 ACC was formed for operations in Palestine and Transjordan, its depot being at Sarafand, Palestine, while in the following year, on 1 December 1923, No 1 ACC was disbanded in order to form an AC Wing at Hinaidi, followed by the establishment of Nos 3, 4, 5 and 6 ACC from the disbanded unit's vehicles and personnel.

Over the next few years further reorganisations of the ACC saw some disbanded and others amalgamated to form a single

AC Wing, Iraq Command, manned by 24 officers and some 200 non-commissioned airmen; the wing now being divided into Sections, with No 1 at Basra, Nos 2-5 at Hinaidi, No 6 at Kirkuk, and No 8 at Mosul.

This latest reshuffle took place from 1 April 1927, just in time to participate in intense active operations along the Iraq-Najd border in 1927-28. During the 1920s the bulk of the AC Wing's vehicles were pre-1920 Rolls-Royce private cars, gutted and rebuilt, with four tons of armour and necessary modifications for their new role, but retaining 45hp Rolls-Royce engines on 'Silver Ghost' chassis. Each car had a single Vickers .0303-inch machine gun mounted in the top turret, plus a normal aircraft compass on a brass standard nine inches high on top of the gun turret for navigation.

A normal fighting crew was three men, fuel was carried for a 350-miles range, along with water and rations for seven days. Other impedimenta included wireless equipment, spares and cans of petrol. Accompanying Ford cars with open box bodies were usually fitted with a Lewis machine gun mounted on an aircraft Scarff Ring. Painted in dull, light khaki outside to merge with the surrounding desert, the Rolls-Royce armoured cars normally had individual names painted on each side of the bonnet e.g. *Cleopatra, Chatham, Cheetah, Avenger, Golden Eagle, Erberus, Hawk, Panther, et al*; selection of names being the prerogative of the unit commander in question. Such semi-official titles were often 'legalised' by the prefix 'HMAC' (His Majesty's Armoured Car) after the naval fashion.

In hostile country the armoured cars usually moved in a pattern consisting of three Ford scout cars about 1000 yards ahead of the main column of armoured cars which, led by the Flag (commander's) AC, drove in line-ahead formation. When any engagement with an enemy was imminent the main column could either spread into a V-formation (like aircraft) or continue in line-ahead and turn (on signal from the Flag car), firing, when within range. When accompanied by aircraft – which normally flew ahead and on the flanks of the main AC column – constant wireless contact between cars and aircraft was maintained. Ideally, such a combined car/aircraft

(*Right*) The official badge of No.1 RAF Armoured Car Company, its motif (appropriately) being a Persian battle chariot.

(*Below*) HMAV *MEDWAY*, a Morris armoured Scout (15hp) designed ca. 1931, with Scarff Ring and Lewis gun mounting on cabin roof.

formation would have the aircraft locate, attack and disperse the rebel tribesmen, leaving the armoured cars to the round up of isolated rebel parties, arrest their leaders, and bring these back for custody, or in the case of trans-border raiders, to chase them back to their own territories.

Much of the armoured cars' duties concerned visiting outlying posts and guarding the overland mail route through Iraq, but also included plentiful surveys of the deserts. These latter duties entailed being away from base for lengthy periods and the car crews were then supplied with essentials by air en route. Navigation, often over bald stretches of desert and scrubland of 200-300 miles, was entirely by compass, and actual progress depended upon the type of surface encountered; speeds of 60 mph being possible on stony or 'steppe' surfaces.

Examples of the car crews' more active operations occurred in 1926-1927 when a 200-strong tribal raiding party from Syria travelled 600 miles towards Koweit (Kuwait), looting camels etc from other tribes along the way over several months. Aircraft from No 84 Squadron located and attacked the main raiding party and dispersed it into smaller groups. A Section of armoured cars following up travelled 250 miles, intercepted the plunderers, killed 29 and captured 26 in a fierce engagement. Early in 1927 another raiding party crossed the border into Iraq. The armoured car section despatched made a 'forced march' of 250 miles in 18 hours and occupied a wadi north of Rutbah Wells. Next morning the cars faced the raiders and the unit commander fired a single shot into the ground in front of the tribesmen – who promptly surrendered.

Once Iraq had attained its 'independence' the RAF garrison forces were reduced and re-organised, including the armoured car companies who were then mainly based in Palestine and Egypt, primarily to assist the army, RAF and civil police authorities in preserving the peace between Arabs and Jews in the Holy Land – a thankless task which brought them into action on myriad occasions during the 1930s.

The original Rolls-Royce cars were gradually replaced or supplemented by more modern armoured vehicles such as

HMAC *JAGUAR* – Rolls-Royce 40-50hp *Silver Ghost* armoured car of No.1 ACC, Mosul, 1935.

HMAC *EAGLE* of No.1 ACC, attached to No.14 Squadron RAF, viewed on 9 May 1923.

Lancias, Alvis, and Crossley 'Six-by-Fours', but Rolls-Royces were to see further operations in North Africa during World War Two, some still bearing their pre-war titles on their bonnets, and were only officially declared obsolete for operations in early 1942, being finally withdrawn and scrapped two years later. The only armoured car unit of the RAF to see regular service beyond the countries bordering the Mediterranean was the Section transferred from Iraq in 1928, which went to Aden where it became titled 'D Flight', No 8 Squadron based at Khormaksar.

For those who flew over the vast wastelands and hill countries of Mesopotamia, Iraq, Persia and India, carrying out routine aerial reconnaissances and surveys, there was often the compensation of being witness to astonishingly beautiful landscapes and raw nature at its most impressive array. Squadron Leader C.H. Keith (later Group Captain), serving on No 70 Squadron in early 1927, was a passenger in a Bristol F2B of No 6 Squadron, based at Mosul in March of that year, and described his view of the mountain range along the Mosul-Turkish border area;

> I set off in a Bristol Fighter piloted by Squadron Leader C.N. Lowe* to do a frontier 'recco'. Snow lay heavy on the mountains and I was wisely wrapped up in all I could muster under my Sidcot flying suit. We set out nor'-east, climbing steeply, but never steep enough, it seemed, as we faced a towering wall of snow. Distance is deceptive, and by the time we reached the foothills we cleared them comfortably at 6,500 feet. Lowe is an old hand amongst these mountains, and he wended his way to and fro but even so we had to get up to 10,700 feet to clear some peaks.
>
> The beauty of those grim, snow-covered mountains makes one gasp. A bright sun shining from a sky, cloudless except for a few wisps around the higher peaks, and yet the wind fits your face like an ice mask. Row after row of misty white peaks, looking like sleeping giants; dark, deep valleys, whose

* Later Group Captain, MC,DFC.

> sides are so sheer that one wonders that even light cares to penetrate into their depths; angry little silvery streams, working themselves into white fury as they twist and turn in their relentless quest for the freedom of the sea; gaunt rocky crags, raising their ugly heads up at us as though inviting our frail, beautiful 'Brisfit' to risk a throw with them ... Dotted in this country are tiny Iraqi police posts, placed there to bring the country under the civil administration and prevent tribes indulging in their pet hobby of raiding and looting each other. No 6 Squadron regularly flies over these posts, and they each display a ground sign code letter which tells the aircraft of their needs, which are at once wirelessed to police headquarters at Mosul, via the RAF ground station. I admire the 'guts' of those mountain police, living their dangerous, lonely life, often completely cut off from the outside world during months of the winter. One post, Ora, lies at the end of a cul-de-sac valley, and one has to execute an alarmingly steep turn to get out again, for the 'Brisfit' could not climb high enough to scale the far end.*

Whether in Iraq or India, flying and operating over and through mountains seemingly reaching to the sun was an awe-inspiring experience which never lessened even after literally years of such activities. Diving into a seemingly bottomless valley between towering stone-faced sheer mountain sides – 'Shooting the nullah' as this was commonly known in India – and never *quite* sure if one's engine might not falter as the pilot began his pull-out to avoid crashing into yet another mountain wall ahead kept the adrenalin pumping hard and fast in any crew's veins. Add to that the common hazard of being shot by tribesmen perched precariously on ledges along the mountain sides, sometimes at the aircraft's level, and – as one pilot expressed his opinion – you 'earned your overseas allowance' ...

Even over less 'elevated' territories flying was beset by virtually every form of weather conditions; massive thunder-

* *Flying Years* by C.H. Keith: J. Hamilton, 1936

storms framed by shimmering lightning, sand storms of blinding fury, heat 'bumps' which could tear the control column out of a pilot's grip and toss an aeroplane up or down a thousand feet in seconds like a limp butterfly. Other potential disasters lurked in the air for the unwary; suddenly plunging through a dense cloud comprised of multi-million locusts on the move to new pastures which painted windscreens and wings with the bright green of their blood on collision, or desert hawks with wing-spans of 15 feet or more whose impact could almost tear a wing off any Ninak or 'Brisfit' head-on, and on occasion even a fatal crash.

Peter Wilkinson flew DH9As in Iraq in the later 1920s and has none-too-fragrant memories of his time there:

> Before my posting to Iraq I'd never been further than Brighton for any holiday, so I rather looked forward to the idea of spending a couple of years in sunny climes – and being paid for it too! My flying experience in England at that time had been on simple aircraft like the Avro 504 and suchlike and I had less than a hundred hours in my log book. I sailed from Southampton on a troopship crowded to the gunwales with a large draft of airmen and army cavalry – complete with stinking horses and being a very junior Pilot Officer was soon detailed by the army OC Troops for various duties aboard, including 'Welfare Officer', whatever that was supposed to be. Visits to the lower decks each day were like descending into some medieval dungeons. Hammocks were slung cheek by jowl, bumping each other, and there was a constant awful smell of sea-sick and unwashed bodies which you could cut with a knife. It must have taken a strong stomach to survive weeks of living in such an atmosphere and the poor airmen had all my sympathies. I eventually arrived at Hinaidi where I was put on the Station strength as a 'supernumerary' pilot instead of joining a proper squadron.
>
> When I arrived Hinaidi was the home of Nos 30, 55 and 70 Squadrons and a busy airfield. 30 and 55 were flying Ninaks and 70 had twin-engined Victoria transports. I was given an office in SHQ as a sort of assistant to the Adjutant and was

not particularly pleased to be regarded as just another pen-pusher, but being a very young officer held my tongue for the time being. It seems the SHQ staff were short of officers due to sickness and other reasons whereas the squadrons had sufficient pilots. After a few weeks of shuffling papers all day I asked the Adjutant if there was any chance of my flying – after all, I *was* a pilot! He was a decent sort and must have had words in the right ears because a few days later he told me that 55 Squadron would let me fly with them if I wished. It would only be occasionally but I jumped at the chance, of course.

My first flight with 55 was as a back-seat passenger. The idea was to show me the sort of territory they regularly patrolled first before they entrusted me with a Ninak and my own back-seat airman. My first impressions of Iraq from the air was to wonder how the hell anyone found their way about – the country was featureless to my ignorant eyes! Navigation in the air had not been my best subject during flying training and I couldn't believe it was possible to do so accurately over these deserts and hills. My pilot, an old hand out there, reassured me that I'd 'soon pick it up', and on my next flight, still a passenger, he went to great lengths to point out salient landmarks and features to remember for guidance. Over the next few weeks I managed to scrounge a couple more 'recce' flights in the back seats of Ninaks, in between my desk duties, then came the great day the 55 chap offered to let me fly a Ninak with him behind me to make sure I didn't get lost. After a briefing on the Ninak's controls, instruments and general handling, I climbed in the front seat nervously – I'd never flown a 9A before – then taxied out to start. Take-off was bloody slow to my mind and the engine seemed to be straining its guts out to gain height in the hot thin air, but we got away without any trouble and climbed to 8,000 feet eventually – it seemed to take hours – before settling down for the actual recce. Guided by my mentor in the back cockpit, I flew around the patrol area steadily, trying desperately to find some of the ground marks he'd pointed out to me before – and failing – until he thumped my shoulder, pointed at the petrol gauge, and indicated it was time to go back to Hinaidi. Coming in to land the Ninak floated and wallowed as heat waves hit it underneath from the surface but I managed to put her down without breaking anything. My 'teacher' seemed happy with my handling of the Ninak but suggested I swot up on those landmarks etc.

A few more flights later I was invited to take part in an actual operation, bombs and all! Very excited at the prospect I went over to the squadron's hangar to the Ninak allocated to me, where I was met by my allotted air gunner; a tall LAC wearing a gunner's winged bullet on his sleeve who was obviously well-versed in 'nursing' young pilots. As he donned a Sidcot suit over his scarves and jumpers – it got bloody cold up there in the back seat as I'd already discovered – he quietly briefed me about the forthcoming flight in terms of how he'd take over any actual bombing runs, and what to do when he wanted to use the Lewis gun. There were three Ninaks on this sortie and we took off in a Vic-formation – a bit of swank – with me struggling to keep my place. Once at cruising height, however, we separated into a bigger gap between aircraft – probably because the leader didn't like the idea of me being *too* close.

Our target was a village at the base of some pretty tall hills at the end of a valley, awkward to get at. I don't remember why we were bombing it. Presumably the villagers hadn't paid a fine, or had been stealing cattle from other tribes, and had ignored the usual government warning notices to 'stop it or else'. Since I didn't need to worry about navigating – just to keep the other Ninaks in sight and follow them – I concentrated on my flying. A sudden burst of gun fire right behind me frightened the life out of me, but on turning my head to see what was going on I was met with a cheeky grin from the LAC and a thumb-up gesture. He'd merely been testing the Lewis to make sure it hadn't frozen up. We soon arrived over the foothills where our target lay and the leader waved one arm around in a 'follow me' gesture. I let the other two aircraft go first, then started my own dive towards the village in their wake.

Flattening out on the approach I then became just a chauffeur as the gunner began sorting out the bombing run. My job was to hold the Ninak rock-steady if I could as we approached, then get the hell out of the valley after the bombs had gone before we hit the hill at the far end. We were carrying 20lb and 112lb bombs, and on the first run-in the LAC dropped the 20-pounders as 'sighters' only. As soon as these were released he reached over and banged my shoulder; the pre-arranged signal to get out of the valley. I promptly yanked the stick back and over with full throttle – that end wall looked awfully close! Staggering clear, I circled and began the second run, aiming for a tall watch tower in the centre of the village. Once again the LAC took care of dropping the 112-pounders, thumped my shoulder, and we climbed away over the nearby hill.

All had gone well up to now and I felt pretty bucked with myself. I could see the other two Ninaks still circling the village and making low level runs, their air gunners firing their Lewis guns at some caves in the hills, presumably inhabited by the evacuated villagers. My gunner indicated that I should follow suit, so down I went for a third time, close to the hillside. The LAC banged away at the cave entrances in short bursts while I kept an anxious eye on my wing-tip and the other eye on the hill ahead, at the same time praying the LAC didn't get too enthusiastic and blow the Ninak's tail off! I was vaguely aware of some unfamiliar popping noises as we swept past the caves but was too busy concentrating on flying to worry about them then. Regaining height I breathed easily again, then looked for the other Ninaks and found them waiting for me to rejoin.

As we flew back to base my gunner tapped me on my helmet. Looking round I found him grinning broadly and pointing at the side of the fuselage just between our cockpits, where there were several neat round holes, and a few more which he pointed to through the starboard lower wing. Bullet holes – which explained those popping noises I'd heard! Back at Hinaidi we landed separately – no fancy formation stuff – and I came in very carefully, keeping in mind my 'teacher's' advice that any sort of side-slip landing would probably wipe off a Ninak's undercarriage. Once down the LAC and I inspected the Ninak for damage. Those fuselage holes had been too close for comfort, only inches away from our bodies. I now believed the stories I'd heard about Arab marksmen …

I managed to scrounge several more flights on operations with 55, most of which were much the same as my first, but was then posted to Heliopolis in Egypt. Although I was disappointed not to have joined a squadron properly in Iraq, I certainly wasn't sorry to leave that benighted country for the fleshpots of Cairo. Iraq in those days was a sink-hole of a posting for any RAF man. The climate was possibly the worst part with temperatures well over a hundred in the daytime and not much lower at night. Sunstroke was a serious menace if you weren't careful. Diseases were rampant among the local population and easily contracted if you didn't take all precautions. Food generally wasn't all that good, while accommodation was little better than the native mud-huts except ours looked prettier with a coat of coloured whitewash. Even in my fairly short time in Iraq I knew of several men – officers and airmen – who went 'round the bend' due to conditions there. It was only just bearable for officers, but the ordinary airmen were worst off. Many of the NCOs in Iraq in my day were of the old school – pre-World War One soldiers who had transferred to the Royal Flying Corps during that war and signed on with the RAF after. Their ideas of discipline must have made life a misery for most airmen.

The best of the bunch were the tradesmen – fitters and riggers etc – who were relatively better educated and treated their own airmen like human beings. Mind you, some of the more senior officers I came across were not happy acquaintances. The RAF then was very class-conscious in the social sense, with a huge gap between commissioned and non-commissioned ranks. Only the bulk of the junior officers – mostly the air crews – were able to mix easily with the 'troops', as we called the Other Ranks. This was due, of course, not just to the officers' youth but mainly because of the flying side, where virtually all air gunners were airmen of lowly rank who shared the pilots' dangers, while the ground crews of each aircraft were literally dedicated to their jobs.*

Wherever the RAF operated overseas pre-1939 a constant corollary to Service life was the deaths of close friends and fellow unit members, both air and ground crews. The causes were many, of course, ranging from 'natural' to disease, flying accidents, operational casualties. Of the latter the fates of most air crews were at least known, but no few simply disappeared as captives of tribesmen who tortured, mutilated and executed them, their final resting place undiscovered or marked. Stemming from a tradition born within the British flying Services during 1914-18, the RAF bore its losses with equanimity, albeit private grief. Death was recognised by all Servicemen as a probable consequence of their calling, but not one to dwell on unduly. During World War One Hugh

* Correspondence with author, 1970.

Typical vulture hazard after an air collision. Note damage to aircraft's lower wing.

Trenchard had always advocated a policy of filling quickly any empty chair in his squadrons' messes resulting from casualties by arranging for immediate replacements. This tradition has continued in the RAF until the present day. Nevertheless, the deaths of fellow crews among the few overseas' squadrons during 1918-38 were particularly felt, epitomising the very close community atmosphere of each individual unit.

That feeling is perhaps well described by the late C. Hilton Keith when he was a commander of No 6 Squadron at Mosul, Iraq in 1928 and described a funeral for two pilots of his unit in a letter dated 8 June 1928:

> I opened my eyes to see George Russell standing near me. 'Sergeant Garner and Sergeant Futcher have spun into the ground, and are both dead,' he was saying. I woke right up then, and swallowing down the nausea which a crash always brings me, I rushed into my clothes. The locality of the crash was in wild country, and quite unapproachable by road transport. I sent out three 'Brisfits' with my MO [Medical Officer] to examine the crash, burn it and bring in the bodies to the Station mortuary.
>
> At noon I heard them overhead, and silently they came to earth, two of them bearing rigid, white-swathed figures, seated in the air gunners' seats – the last flight those two poor fellows would ever make. The heat was terrific and MacWeeney [MO] when he got back looked pretty grim. I told him to keep quiet and my second MO went on with the sad preparations, and in the pitiless sun two of my coolies started digging graves in the little British cemetery, just outside the old walls of Mosul, and another party started to line two waiting coffins with pitch. At 2 pm the lids were screwed down in place, and at 6 pm the funeral procession was lined up and ready to leave. I ought to have read the funeral service myself, but I got Flight Lieutenant Craik, who has relieved Stedman in charge of the armoured cars, to take it, as he once used to be a parson. Flowers came in pathetic profusion from all sections of Mosul – came to be carried on the RAF ensigns which draped the coffins, and then to give

up their sweetness and wither as they lay on the sun-scorched ground of that shadeless, sad little cemetery. The crowds at the cemetery spoke well of the good feeling which exists towards my people. The Commandant of the [Iraq] Levies came with all his officers, and we silently filed through a party of Levies he had sent, standing on either side of the approach road with their brown heads bowed on their reversed rifles – Assyrian hill-fighters, yet Christians, like ourselves.

Charles had sent a detachment of police and the Iraqi Army Commandant came with many officers. The service seemed never-ending, and as one stood to attention in the breathless violent heat of the dying sun, myriads of flies buzzed around our heads. Then blank volleys, 'The Last Post' and the 'Reveille', and we all filed past the open graves to salute all that was left of those poor battered bodies which, not twelve hours ago, had been cheery athletic fellow-countrymen, full of joy and life. I have been busy writing to their 'next of kin' – mercifully neither was married.*

The two sergeants referred to by Keith were but representative of several hundreds of RAF men who left their native shores and never returned. Each, in his own modest way, had fulfilled his duty to his Service and to his country, culminating in the ultimate sacrifice. Too many now lie in forgotten graves, victims of history, names on a lengthy Roll of Honour. For them the words of Rupert Brooke are especially appropriate:

If I should die, think only this of me:
That there's some corner of a foreign field
That is forever England.

* *Flying Years* by C.H. Keith; J. Hamilton, 1936

CHAPTER THIRTEEN

On the Brink

The events in Europe during the year 1938 which culminated, in a sense, in the Anglo-German Munich agreement of 29 September are well recorded in the contexts of political and diplomatic manoeuvrings by Hitler in his relentless pursuit of yet more *Lebensraum* for his Nazi empire. In Britain this Munich appeasement – which calmly sacrificed Czechoslovakia in the cause of 'peace for our time' as the British Prime Minister Neville Chamberlain naively expressed it – proved to be merely a temporary relief; a breathing space for one more year in which to continue its hasty re-armament programmes for the inevitable war to come. At the moment of the Munich agreement the RAF possessed 149 firstline squadrons in total, with several others about to be reformed, and of these 29 were based overseas, as follows:

Squadron	*Main base*	*Aircraft*
5	Risalpur, India	Westland Wapiti II
6	Ramleh, Palestine	Hawker Hardy
8	Khormaksar, Aden	Vickers Vincent
11	Risalpur, India	Bristol Blenheim I
14	Amman, Transjordan	Vickers Wellesley
20	Peshawar, India	Hawker Audax
27	Kohat, India	Westland Wapiti II
28	Ambala, India	Hawker Audax
30	Habbaniya, Iraq	Bristol Blenheim I
31	Drigh Road, India	Westland Wapiti II
33	Heliopolis, Egypt	Gloster Gladiator
36	Seletar, Singapore	Vickers Vildebeest
39	Risalpur, India	Hawker Hart
45	Helwan, Egypt	Vickers Wellesley

Squadron	*Main base*	*Aircraft*
47	Khartoum, Sudan	Vickers Vincent
55	Habbaniya, Iraq	Vickers Vincent
60	Kohat, India	Westland Wapiti II
70	Habbaniya, Iraq	Vickers Valentia
84	Shaibah, Iraq	Vickers Vincent
100	Seletar, Singapore	Vickers Vildebeest
113	Heliopolis, Egypt	Hawker Hind
202	Alexandria, Egypt	Saro London
203	Basra, Iraq	Short Singapore
205	Seletar, Singapore	Short Singapore
208	Mersa Matruh, Egypt	Hawker Audax
211	Helwan, Egypt	Hawker Hind
216	Heliopolis, Egypt	Vickers Valentia
223	Nairobi, Kenya	Vickers Wellesley
230	Seletar, Singapore	Short Sunderland I

As will be immediately apparent, only six of these 29 squadrons were flying modern monoplane aircraft, although several units would exchange their outdated biplanes for newer designs over the following year. As always, the overseas squadrons were last in priority for fresh equipment.

Although actual mobilisation of the RAF was not *officially* instigated in September 1938, all Commands were ordered to prepare for same, and certain emergency measures put into hand. These measures included all overseas' squadrons who were expected to be ready to implement their long-practised reinforcement moves e.g. some units in India to reinforce the meagre RAF strength at Singapore, while a few units actually moved to their war stations, as in the case of No 202 Squadron which moved from Malta to Alexandria temporarily, then returned to Kalafrana once the crisis had dissolved.

To those men who had been serving overseas for several years by 1938 the European crisis seemed somewhat remote. As one officer described his feelings:

> After three years in India, Britain, Germany and Czechoslovakia seemed too far away to have any impact on my life.

> Although I recognised the seriousness of the situation to a degree, the name Hitler meant very little to most of us, and who had ever visited Czechoslovakia?

It was, perhaps, an excusable viewpoint in that even had war broken out in 1938 in Europe, the men of the overseas units, particularly in India and the Far East, had no immediate enemy to contend with as far as they could see.

In the Middle East zones the RAF tended towards a more serious attitude to the 1938 crisis. In and around the Mediterranean, such strategic points as Gibraltar, Malta, and the Suez Canal were vital to British interests, while the oil lands east of the 'Med' were ripe for attack should war be precipitated. If Hitler was a vague menace to some Servicemen in the Middle East, the egotistical and wholly unpredictable dictator of Fascist Italy, Benito Mussolini, was very much a clear and potential menace. Italian empire-building along the North African coastal areas, added to the recent invasion and conquest of Abyssinia, had already given Mussolini's ego an inflated opinion of his 'destiny', and might well have led to further excesses, particularly in view of his alliance with the German dictator, Hitler.

To most Britons the prime safeguard remained the Royal Navy's ability to 'rule the waves' anywhere in British Empire (or other) waters; an ingrained traditional trust in naval power which, even by 1938, made no concession to the potential of air power. Yet, as that champion of air power, the late Group Officer C.M. McAlery, WRAF, exclaimed in a survey of the Empire's air defences in *The Aeroplane*, 21 September 1938:

> The Navy cannot defend the Suez Canal without air co-operation, it can no longer protect our interests in the Mediterranean, it is helpless against an enemy air force based in Africa, it cannot prevent the invasion of India and it cannot even protect its own great base at Singapore. Without shore-based aeroplanes at strategic points the Fleet cannot keep our lines of communication free or find safety in its own harbours.

Hawker Hinds of No.113 Squadron shortly after arriving in Egypt, 1938.

Short Singapore III, K3594 of No.205 Squadron, based at RAF Seletar, Singapore. The heart insigne was painted red.

Biased words perhaps, but prophetic ...

By 1 October 1938 the semi-mobilisation of Britain's forces had been relaxed and all units returned to their normal bases and duties. For most squadron crews on overseas' units the crisis had meant very little in the context of prospective war operations, but had emphasised to the more thoughtful the obsolescence of their aircraft had such operations been required against any opposing air force. In Britain the rapid expansion of UK-based units was clearly emphasising the government's near-obsession with defence of the mother country from an expected massive aerial bombing assault immediately upon the outbreak of any war in Europe*, hence British-based squadrons had overwhelming priority for the latest aircraft designs, especially fighters.

Such a decision, albeit logical at first glance, failed to take into account the overall effect on Britain should its ill-armed and ill-defended empire overseas be attacked, conquered, and thereby isolated from the UK. There was also a modicum of irony in this policy of equipping RAF units in Britain with the best available aircraft *et al* to the detriment of overseas' squadrons – few UK-based squadrons had any real experience of actual war operations during 1918-38, whereas all overseas' units during those two decades had rarely experienced *non*-operational flying for any lengthy period.

The Munich crisis, as far as overseas' RAF squadrons were concerned was simply one more temporary burden. Active operations were still being undertaken in various troubled areas, in particular Palestine and Aden, and to a slightly lesser degree India's North-West Frontier Province, as witnessed by the continuing awards of various gallantry decorations to both air and ground crews in 1938-39. The contemporary crews involved in these operations were but the latest of several thousands of men of the RAF who had served, fought, and too often died many leagues from their native lands during the previous 20 years. All had gained unforgettable experiences in

* A fear, shared by France and other European countries, exacerbated by many years of over-emphasis on the destructive power of the bomber, and exaggerated estimates of Hitler's Luftwaffe's capabilities and strength.

Fairey Gordon, K2645 of 47 Squadron, based at Khartoum, near Jebel Kassala, 1933 (pilot, Flg Off J B T Whitehead).

Short Sunderland I, L2160 of No.230 Squadron being 'christened' as *SELANGOR* at Port Swettenham, 19 October 1938 by the Sultan of Selangor.

the hardest 'school' of actual operations against foes ready and willing to kill. Other facets of their service abroad – long distance flights, aerial surveys, mercy missions, air transportation and supply of ground forces and the like – had deepened and widened their knowledge and efficiency. All had been undertaken willingly despite far from ideal conditions, lack of finance and adequate facilities, in aircraft for the most part of near-vintage design. If the RAF in Britain came to be cynically titled as 'the finest flying club in the world', the overseas units could have rightly earned the description 'Shoe-string Air Force'.

Such inculcated experience of the young officers and airmen of the 1920s and early 1930s was, by 1938, providing a thinly-spread hard core of veterans throughout most levels of the RAF at home and abroad. Many such men had by then begun reaching more senior ranks and command posts in the expanding Service, while even the most senior officers at staff levels, many with operational backgrounds stretching from the 1914-18 war, had in many cases in-the-field first-hand experience of operational command.

Thus, when the RAF went to war again in 1939 with Germany it was by no means the unprepared and untested Service it had often been described in certain past published 'histories' and biographies. Certainly, the 1939 RAF lacked adequate overall strength and modernity in aircraft, but its serving personnel were true professionals, well-versed in their individual skills and crafts, needing only the right tools in order to carry out their assigned tasks and duties.

No few of those men were the fruits of Hugh Trenchard's visionary plans for 'his' air force from its birth, men apprenticed to their professions at such foundations as Cranwell, Halton, Flowerdown and Eastchurch. His immense foresight and judgment were to be justified a thousandfold in the years which followed.

APPENDIX I

Squadrons based overseas, 1919-38

NB: Excluding RAF squadrons based in Europe at 1918 Armistice subsequently reduced to cadre and disbanded, and Fleet Air Arm units.
* Main bases, excluding detachments.

Squadron	*Period(s)*	*Bases**	*Aircraft*	*Remarks*
1	Jan '21 – Nov '26	Risalpur Bangalore Hinaidi	Snipe Nighthawk	Reformed Risalpur 21/1/20 To Iraq 27/4/21 Disbanded Hinaidi 1/11/26
2	Apr '27 – Oct '27	Shanghai	Bristol F2B	
3	Feb '20 – Oct '21 Oct '35 – Aug '36	Ambala Khartoum Port Sudan	Snipe Bulldog	Reformed Ambala 1/2/20 Disbanded Ambala 1/10/21 Abyssinian Crisis
4	Sep '22 – Sep '23	Kilia Kilid el Bahr	Bristol F2B	Chanak Crisis

Squadron	*Period(s)*	*Bases*	*Aircraft*	*Remarks*
5	Feb '20 – WW2	Quetta Ambala Dardoni Kohat Tank Risalpur	Bristol F2B Wapiti	48 Sqn renumbered, Quetta 1/2/20
6	Ápr '19 – WW2	Mosul Ismailia	Bristol F2B Gordon Hart	
8	Oct '20 – WW2	Helwan Baghdad Khormaksar	DH9A Fairey IIIF Vincent	
11	Dec '28 – WW2	Risalpur	Wapiti Hart	
12	Oct '35 – Aug '36	Khormaksar Robat	Hart	Abyssinian Crisis
14	Feb '20 – WW2	Ramleh Amman	Bristol F2B DH9A Fairey IIIF Gordon Wellesley	Reformed Ramleh 1/2/20 (111 Sqn renumbered).

Squadron	*Period(s)*	*Bases*	*Aircraft*	*Remarks*
17	Dec '18 – Nov '19	Amberkoi Batum	Camel/DH9	Disbanded San Stefano 14/11/19
20	May '19 – WW2	Risalpur Ambala Quetta Peshawar Miranshah	Bristol F2B Wapiti Audax	
22	Oct '35 – Aug '36	Hal Far	Vildebeest	Abyssinian Crisis
25	Sep '22 – Oct '23	San Stefano	Snipe	
27	Apr '20 – WW2	Mianwali Risalpur Dardoni Miranshah Kohat	DH9A Wapiti	99 Sqn renumbered, Mianwali 1/4/20
28	Feb '20 – WW2	Ambala Kohat Peshawar Quetta Manzai Ambala	Bristol F2B Wapiti Audax	114 Sqn renumbered, Ambala 1/2/20

Squadron	*Period(s)*	*Bases*	*Aircraft*	*Remarks*
29	Oct '35 – Sep '36	Amriya Helwan	Demon	Abyssinian Crisis
30	Feb '20 – WW2	Baghdad West Hinaidi Mosul Dhibban*	RE8/DH9A Wapiti Hardy	Restored from cadre 1/2/20 *Renamed Habbaniya 25/3/38
31	1916 – WW2	Risalpur Maow Cawnpore Peshawar Dardoni Ambala Quetta Karachi	BE2C & various Bristol F2B Wapiti	
33	Oct '35 – WW2	Mersa Matruh Amman Gaza Ismailia Heliopolis	Hart Gladiator	
35	Oct '35 – Aug '36	El Damer Gebeit	Gordons	Abyssinian Crisis
36	Dec '30 – WW2	Seletar	Horsley Vildebeest	

Squadron	*Period(s)*	*Bases*	*Aircraft*	*Remarks*
39	Jan '29 – WW2	Risalpur	Wapiti Hart	
41	Oct '35 – Sep '36	Khormaksar Sheik Othman	Demon	Abyssinian Crisis
45	Apr '21 – WW2	Helwan Almaza Hinaidi Heliopolis Helwan	Vimy Vernon DH9A Fairey IIIF Vincent Hart Gordon Wellesley	Reformed Helwan 1/4/21
47	Feb '20 – WW2	Helwan Khartoum	DH9/DH9A Fairey IIIF Gordon	Reformed Helwan 1/2/20
48	Jun '19 – Mar '20	Quetta	Bristol F2B	Renumbered 5 Sqn at Quetta, 1/4/20
55	Feb '20 – WW2	Suez Maltepe Baghdad West Mosul Hinaidi Dhibban	DH9A Wapiti Vincent	Reformed Suez 1/2/20 (142 Sqn renumbered)

Squadron	*Period(s)*	*Bases*	*Aircraft*	*Remarks*
56	Feb '20 – Sep '22	Aboukir	Camel/Snipe	Reformed Aboukir 1/2/20 (80 Sqn renumbered) Disbanded Aboukir 23/9/22
58	May '19 – Feb '20	Heliopolis	HP 0/400/Vimy	Renumbered 70 Sqn 1/2/20
60	Apr '20 – WW2	Lahore Risalpur Dardoni Miranshah Kohat	DH10 DH9A Wapiti	Reformed Lahore 1/4/20 (97 Sqn renumbered)
63	1918-20	Baghdad	RE8	Disbanded Baghdad 29/2/20
64	Mar '36 – Aug '36	Heliopolis Ismailia	Demon	Reformed Heliopolis 1/3/36 Abyssinian Crisis
65	Jul '35 – Jan '36	Malta	Demon	Part-squadron only for Abyssinian Crisis
70	Feb '20 – WW2	Heliopolis Baghdad West Hinaidi Dhibban	Vimy Vernon Victoria Valentia	Reformed Heliopolis 1/2/20 (58 Sqn renumbered)
74	Sep '35 – Jul '36	Hal Far	Demon	Reformed 3/9/35 en route to Malta

Squadron	*Period(s)*	*Bases*	*Aircraft*	*Remarks*
84	Aug '20 – WW2	Baghdad West Shaibah	DH9A Wapiti Vincent	Reformed Baghdad West 13/8/20
97	Jul '19 – Apr '20	Allahabad Lahore	DH10	Renumbered 60 Sqn 1/4/20 at Lahore
99	May '19 – Apr '20	Ambala Mianwali	DH9A	Renumbered 27 Sqn, Mianwali 1/4/20
100	Sep '33 – WW2	Seletar	Vildebeest	
111	1918 – Feb '20	Ramleh	Bristol F2B	Renumbered 14 Sqn, Ramleh 1/2/20
113	Apr '38 – WW2	Heliopolis	Hind	
114	1917 – Jan '20	Lahore	BE2 & various	Renumbered 28 Sqn, Lahore 31/1/20
142	Oct '35 – Dec '36	Aboukir Mersa Matruh Helwan Ismailia	Hart	Abyssinian Crisis

Squadron	*Period(s)*	*Bases*	*Aircraft*	*Remarks*
202	Apr '20 – May '21	Alexandria	Fairey IIID	Reformed Alexandria 9/4/20 & disbanded 16/5/21.
	Jan '29 – WW2	Kalafrana	Fairey IIIF Scapa London	Reformed Kalafrana 1/1/29
203	Feb '29 – WW2	Aden Basra	Southampton Rangoon Singapore	
204	Sep '35 – Aug '36	Aboukir Alexandria	Scapa	Abyssinian Crisis
205	Jan '29 – WW2	Seletar	Short Singapore	Reformed Seletar 8/1/29 from the Far East Flight.
207	Sep '22 – Sep '23 Oct '35 – Sep '36	San Stefano El Damer Gebeit	DH9A Gordon Wellesley Fairey Battle	Chanak Crisis Abyssinian Crisis
208	Feb '20 – WW2	San Stefano Ismailia Heliopolis Mersa Matruh Heliopolis	Bristol F2B AW Atlas Audax	Reformed Ismailia 1/2/20 (113 Sqn renumbered).

Squadron	*Period(s)*	*Bases*	*Aircraft*	*Remarks*
211	May '38 – WW2	Helwan Ramleh	Hind	
216	Jul '19 – WW2	Kantara Heliopolis	HP 0/400 DH10 Vimy Victoria	
221	1918 – Sep '19	Mudros Baku Petrovsk	DH9/DH9A	Disbanded 1/9/19
223	Dec '36 – WW2	Nairobi	Wellesley	
230	Sep '35 – Aug '36 Oct '36 – WW2	Alexandria Seletar	Singapore Sunderland	Abyssinian Crisis

APPENDIX 2

RAF Songs and Verses, 1918-38

A prominent biographer once contended that if he knew a people's songs he could write their history. In the context of the Royal Air Force, and its predecessors, such an assertion remains true. Like all British Servicemen, RAF men and women have always lightened their trials and tribulations with an esoteric brand of humour, a twisted humour, not readily comprehensible to those listeners or readers lacking Service experience. Mock-cynical, overtly fatalistic, earthily pragmatic, RAF humour may well baffle a modern psychologist with its constant content of such matters as death, injury, even mutilation, yet such subjects are really of no more importance to the regular-serving Serviceman than a cancelled leave pass or disciplinary charge in the long view.

Such songs provided a slim barrier against the full impact of fatalities, injuries, hardships and privations; an outlet for suppressed emotions and a safety valve for mounting stress when needed. All such songs and verses were, quite naturally, sprinkled with pure RAF 'slanguage' with its myriad mixture of foreign language words and phrases accumulated over the years of overseas' service in many lands, and equally naturally include a selection of vulgarities and swear words sounding harsh, even repulsive to the civilian ear. Such language – usually termed 'barrack-room' by the unknowing – has little meaning to the airman, being simply a more emphatic method of making his point. Nevertheless, Kipling's words should be borne in mind before condemning such behaviour: 'Why, single men in barricks don't grow into plaster saints ...'

To set down even a fair proportion of all the RAF songs and verses of the Service's 70 years existence would require several volumes of their own. One such attempted compendium commended to the reader is *Airman's Song Book* by C.H. Ward-Jackson; Sylvan Press, 1945, although most of the famous songs within were 'deodorised' for the contemporary public. In this book I have selected merely three traditional songs to exemplify all others, though in my own experience I consider these particular examples to have been the best-known and most frequently sung at RAF alcoholic-cum-celebratory occasions overseas. I have also retained the 'original' wording as far as possible in order to give the authentic flavour of such duties. Readers with 'sensitive' ears may be partly mollified when

I point out that these three songs are almost respectable compared with some I might have chosen which, even in the present uncensored era of printed language, are best left to memory and private reunions ...

* * *

'THE SOMERSETSHIRE'

Not the English county but the queen of pre-1939 troopships which regularly conveyed overseas drafts of Servicemen to and from their next postings abroad. A ship of the Bibby Line.

We're leaving Khartoum
By the light of the moon
We're sailing by night and by day
We've just passed Kasfareet*
We've had fuck-all to eat
'Cos we've thrown all our rations away'

Chorus:

Shire, Shire, Somersetshire
The skipper looks on her with pride
He'd have a blue fit
If he saw any shit
On the side of the Somersetshire.
Oh, this is my story
This is my song
I've been in this Air Force
Too fucking long
So roll on the Nelson, the Rodney, the Hood
For this fucking dump is just no fucking good.'

NB: The last two lines were 'Updated' during World War Two to:
'Roll on the Nelson, the Rodney, Renown,
You can't sink the Hood 'cos the bastard's gone down'.

SHAIBAH BLUES

RAF Shaibah (pronounced 'Shy-Buh'), located some ten miles south-west of Basra in the Persian Gulf, at the delta of the rivers Tigris and Euphrates, was regarded by most airmen of the pre-1939 RAF as the 'furthest-flung outpost of the Empire', although worse postings in terms of remoteness from 'civilisation' can be readily

* Kasfareet was an RAF station on the Suez Canal banks.

recalled by some. The only RAF unit to be permanently based there pre-1939 was No. 84 Squadron – from September 1920 until September 1940 – although several units sent detachments there for operations in 1941. Thus, if anyone can lay claim to the SHAIBAH BLUES as 'their' song, 84 Squadron has priority.

'Oh, a little bit of *mhutti** fell from out the sky one day
And it fell into the desert fucking miles and miles away
And when old Trenchard saw it,
It looked so bloody bare,
He said "That's what I'm looking for,
I'll put my air force there".
So he sent out river gunboats, armoured cars and SHQ
And then shoved 84 right out there in the fucking blue.
But *maleesh*†, soon I'm going to a land that's far remote
Until that day, you'll hear me say
ROLL ON THAT FUCKING BOAT!'

Chorus:
'I've got those Shaibah Blues, Shaibah Blues
I'm fed-up, I'm fucked-up, and I'm blue
I've tried to learn the lingo but its fairly got my goat
The only thing that I can say's
ROLL ON THAT FUCKING BOAT!'

NO BALLS AT ALL

This particular gem, often titled SULAIMAN, referred to a particular phase of operations against a Kurdish revolt around the Sulaiman area led by the notorious Sheikh Mahmud in 1924, and its prime theme was the probable fate of any airman falling into the hands of rebel tribesmen – especially their women. Only the main (of many) verses are quoted here – others varied over subsequent years but usually gloried in pure physical details ...

'There once were two pilots who went to bomb Sul
Their bombs were OK but their tanks were half-full
Then came from the back seats an agonised call
"If the engine cuts out we'll have no balls at all"

Chorus:
'No balls at all, No balls at all
If the engine cuts out we'll have no balls at all'

* *Mhutti* – much, dirt (to put it politely ...)
† *Maleesh* – never mind.

Twas just over Sul that both engines cut out
And again from the back came a terrified shout
"If we land to the east of the Basrian Pass
Might as well stuff the Lewis up your arse"
Chorus: Etc.
'They looked o'er the side and there plain to see
Sheikh Mahmud and cronies all sat down to tea
A'sitting around midst their goats and the rocks
Discussing spring fashions, long knives and cocks
Chorus: Etc.
'They landed, then ran like chaff 'fore the wind
With fucking great knives flashing inches behind
They knew they were in for 'orrible thrills
So clutched their balls and ran for the hills'
Chorus: Etc.
'The moral of all this is quite plain to see
Fill up your fuel tanks where'er you may be
And if among Arabs and Kurds you must roam
Make sure there's enough to get you back home'
Chorus (sung falsetto)
'*NO balls at all, NO balls at all*
If the engine cuts out you'll have no balls at all'.

APPENDIX 3

RAF Slanguage

British Servicemen serving overseas in the many years prior to 1939 seldom felt it necessary to learn any local native language, relying instead upon a limited vocabulary of the most common, useful words and phrases to be interjected within their normal speech. Over those years (and indeed for many years thereafter) such snippets of varying foreign tongues became common parlance in Service barracks and workplaces, though usually 'anglicised' for convenience. The following selection from such 'slanguage' reflects not only the most commonly used words but also the widely varying derivations. Spelling of these here is deliberately 'Service' phonetical.

Ackers Money (Arabic origin)
Alakeefik Easy-going, couldn't care less (Arabic)
Astee Take it easy, slowly (Indian)
*Badmash*Native Indian robber-burglar/bandit

Baksheesh Money, tip, free handout (Arabic)
Bandabust Deal, arrangement (Indian)
Bhisti Native Indian water-carrier e.g. Gunga Din
Bint Female (Arabic)
Bistra Service bed-roll (Indian)
Blood waggon Ambulance
Bobbajee Cook (Indian, from *bawachi*)
Bolo To speak (Indian) e.g. 'He bolos the lingo'
Brat Fully, Trenchard Brat i.e. ex-Aircraft Apprentice
Bukshee Free, gratis, surplus (Indian)
Bund Perimeter wall/fence around camp (Persian)
Bundoo Differently applied to desert or rifle
Char/Chai Tea (drink) – various derivations
Charpoy Bed of local manufacture (Indian)
Chatti Large earthenware vessel holding cool water
Chico Baby, small boy or girl
Chiefy Flight Sergeant
Chippy Carpenter tradesman
Chota Little, small e.g. 'Chota Peg' – small whisky
Chowkidar Indian tribal village watchman
Coggage Paper (Indian, from *kaghaz*)
Conner Food
Cooler Detention gaol-cell; alts. 'Jug' & 'Mush'
Cooloo All, everything, the lot
Darzi Indian native tailor
Dekko Look, see (Indian, from *dekna*); see 'Shufti'
Dohee Laundry (Arabic, from *dhobi*)
Doolally Mad, insane (from Deolali reception camp, India)
Erk Any airman below rank of Corporal
Feloose Money (Arabic)
Felucca Arab sailing boat
Fizzer Disciplinary charge; alt. 'On the peg'
Gash Spare, surplus, free. See 'Bukshee'
Gharry Horse-drawn open carriage
Gippo Corruption of Egyptian; used as noun and adjective
Golden Eagle Day Pay day (Fridays)
Gong Medal of any sort
Goolie Testicle; also alt. 'Clanger', 'Pill', 'Nut'
Iggeri Hurry up (Arabic)
Imshee Go away! Buzz off! (Arabic)
Irons Knife, fork and spoon eating utensils, Service issue
Jankers Term of disciplinary punishment
Jirgah Tribal meeting/convention (Indian)
Joystick Aircraft/pilot's control column
Lackery Literally wood (Indian *lakri*); any 'thick' airman

Maknoon Mad, out of his mind (Arabic)
Maleesh Never mind, forget it (Arabic)
Mali Indian native gardener
Malik Indian tribal village head man
Matlo Sailor (French *matelot*); alt. 'Fish-head'
Mucker Pal, friend (Indian *makkhn*)
Munshi Indian native teacher
Muskeen Mean, miserly (Arabic)
Ninak De Havilland 9A aircraft (Nine-Ack')
Oppo Close friend i.e. 'Opposite number'
Parni Water, sea (Indian)
Plumber Originally armourer, later all technical tradesmen
Pongo Any member of the army
Props Cloth propeller arm badge worn by Leading Aircraftmen
Puckaroo Pick up, lift (Indian)
Puggled/paggled Insane, mad, incapable (Indian)
Pukkha Correct, genuine (Indian)
Punkah Ceiling fan, man-operated (Indian)
Quoise Good, excellent (Arabic)
Rooti Medal RAF Long Service and Good Conduct Medal awarded for 18 years unblemished service (Indian deriv.)
ROTB 'Roll On That Boat' – universal cry of the overseas airman
Scribe 'Pen-pusher' i.e. administrative tradesman
Shikari Indian native boat/gondola
Shufti Look, see (Arabic); see 'Dekko'
Shylock Accountant tradesman
Sky pilot Service Padre; alt. 'Bible-puncher' and 'God-Botherer'
Sparks Airman of wireless trade
Strafe Aerial low-level attack (German *strafe*)
Swede Yokel, from agricultural background
Tapes NCOs' rank chevrons worn on tunic/shirt sleeves
Tiffin Mid-day meal
Tonga Horse-drawn, two-seat Indian carriage
Tora Peechi Later on, in a while (Indian)
Two-Six! Universal RAF cry for all available men to help
Two-Five-Two RAF Form number of disciplinary charge sheet
Two-Nine-Five RAF Form number of official leave pass
Twelve-Fifty RAF Form number of personal Identity Card
U/S Unserviceable
U/T Under training, unqualified
Wad Cake, bun
Wallah Man, person (Arabic *walad*)
Wog Any local native; said to derive from 'Westernised Oriental Gentleman'
Zizz Sleep, nap

Associated Literature

Compared with the towering mountain of published literature available concerning RAF men, machines, and operations during the war years 1939-45 and more recent years, any catalogue of published works relevant to the first 20 years of the RAF's existence appears to be a mere hillock. Moreover, a high proportion of the latter titles now are rarely found outside individuals' private collections or official libraries and other archives. Thus the following list is offered as a very basic bibliographical guide to the reader seeking in-depth research sources and first hand descriptions of particular and general facets of RAF operational activities during the Third Service's years of adolescence, 1918-38.

Air Publication 125, 2nd Rev. Ed., Air Ministry, 1936
History of the Air Ministry, C.G. Grey; Allen & Unwin, 1940
Royal Air Force, 1918, C.C.H. Cole; Wm Kimber, 1968
Trenchard, A Boyle; Collins, 1962
The Third Service, P. Joubert; Thames & Hudson, 1955
Empire of the Air, Templewood; Collins, 1957
Wings of Destiny, Londonderry; Macmillan, 1943
The Central Blue, J. Slessor; Cassell, 1956
Swifter Than Eagles, J. Laffin; Blackwood, 1964
From Many Angles, F. Sykes; Harrap, 1942
From Sea to Sky, A. Longmore; Bles, 1946
'Pip' Playfair, Playfair/Jarvis; Stockwell, 1979
Flying Fever, S.F. Vincent; Jarrolds, 1972
A Pride of Unicorns, J. Pudney; Oldbourne, 1960
Flying & Soldiering, R.R. Money; Nicholson & Watson, 1936
Last Train Over Rostov Bridge, Aten/Orrmont; J. Messner, 1961
The Czar's British Squadron, Perrett/Lord; Wm Kimber, 1981
Tedder, R. Owen; Collins, 1952
Survivor's Story G. Gibbs; Hutchinson, 1956
To Know the Sky, P. Hill; Wm Kimber, 1962
The Baghdad Air Mail, R. Hill; Arnold, 1929
Flying Years, C.H. Keith; J. Hamilton, 1936
Wings Over Kabul, Ivelaw-Chapman/Baker; Wm Kimber, 1975
To the Ends of the Air, G. Livock; HMSO, 1973

Over the Way with the RAF, H. Howe; A.H.Stockwell, u/d
When the Middle East was Fun, F.R. Wynne; P. Skelton, u/d
Something New Out of Africa, 'HW'; Pitman, 1934
Fly Past, A.G. Lee; Jarrolds, 1974
The Luck of the Devil, A.G. Dudgeon; Airlife, 1985
Never Stop the Engine When its Hot, D. Lee; Harmsworth, 1983
You Are Not Sparrows, S.J. Carr, Ian Allan, 1975
Stand Easy, B.J. Hurren; J. Long, u/d
An Airfighter's Scrapbook, J.I.T. Jones; Nicholson & Watson, 1938
Undiscovered Ends, J.S. Poole; Cassell, 1957
The RAF, B. Robertson; Hale, 1978
Wheels of the RAF, B. Robertson; PSL, 1983
History of the RAF, C. Bowyer; Hamlyn, 1977
RAF Yearbook, 1938, L. Bridgman; Burge, 1939
Pictorial History of RAF, Vol. 1, J. Taylor/C. Bowyer; Ian Allan, 1980
Aircraft of the RAF Since 1918, O. Thetford; Putnam, 1976
Encyclopaedia British Military Aircraft, C. Bowyer; A & AP, 1982
Bristol F2B Fighter, C. Bowyer; Ian Allan, 1985
But Not in Anger, C. Cole/R. Grant; Ian Allan, 1979
From the Ground Up F. Adkin; Airlife, 1983
The Forgotten Ones, P. Joubert; Hutchinson, 1961
Customs & Traditions of the RAF, P.G. Hering; Gale & Polden, 1961
Dictionary of RAF Slang, E. Partridge; M. Joseph, 1945
Airman's Song Book, Ed. C. Ward-Jackson; Sylvan Press, 1945
Bomber Squadrons of the RAF, P. Moyes; Macdonald, 1964
Coastal & Support Squadrons of RAF, J. Rawlings; Jane's, 1982
Squadrons of the RAF, F. Robertson; FLIGHT, 1935
Bombing Colours, 1914-37, B. Robertson; PSL, 1972
British Military Aircraft Serials, 1911-79, B. Robertson; PSL, 1979
The Royal Air Force T. Sprigg; Isaac Pitman, 1935
History of RAF Changi, H.A. Probert; RAF Changi, 1965
Lion in the Sky, N. Shorrick; Federal Pubs, 1968
Guns in the Sky C. Bowyer; J.M. Dent, 1979
Mission Completed B. Embry; Methuen, 1957
Air Command R. Collishaw/Dodds; Wm Kimber, 1973
Twice Vertical, M. Shaw; Macdonald, 1971
Oculi Exercitus (6 Squadron), C.D. Stewart; RAF, 1964
History of 20 Squadron RAF, N. Robertson; Private, 1987
With the First in the Field (27 Sqn), A.E. Cowton; D. Cowton, 1963
The Flying Elephants (27 Sqn), C. Bowyer; Macdonald, 1972
No. 31 Squadron RAF 31 Squadron, 1960
First in the Indian Skies (31 Sqn), N. Franks; RAF Collection, 1981
The Winged Bomb (39 Sqn), K. Delve; MC Pubs, 1985
Over the Balkans & South Russia (47 Sqn), H.A. Jones; Arnold, 1923
Sixty Squadron, Warne/Young; 60 Sqn, 1967

I Fear No Man (74 Sqn), D. Tidy; Macdonald, 1972
No. 100 Squadron, 1917-77, A.J. Brookes; 100 Sqn, 1978
History of 208 Squadron, D. Marr; Private, 1967
The Day We Almost Bombed Moscow, Dobson/Miller; Hodder & Stoughton, 1986
The Frontier, 1839-1947, J.G. Elliot; Cassell, 1968
North-West Frontier, A. Swinson; Hutchinson, 1967
The North-West Frontier, M. Barthorp; Blandford, 1982
The Pathans, 550 BC-AD 1957, O. Caroe; Macmillan, 1958
Solitary in the Ranks, H.M. Hyde; Constable, 1977

Periodicals/magazines/etc
Royal Air Force Record, 1920-21, Air Ministry, 1920-21
Royal Air Force Quarterly, 1930-38, Ed. C. Burge; Gale & Polden
FLIGHT, 1918-38
AEROPLANE, 1918-38
AIRCRAFT ILLUSTRATED EXTRA No. 6, Ian Allan, 1973
Vickers Vimy, J.M. Bruce; *PROFILE* No. 5
Westland Wapiti, C. Andrews, *PROFILE* No. 32
Fairey IIIF, F. Mason; *PROFILE* No. 44
Sopwith Snipe, J.M. Bruce; *PROFILE* No. 50
Hawker Hart, F. Mason; *PROFILE* No. 57
Bristol Blenheim I, P. Moyes; *PROFILE* No. 93
Gloster Gladiator, F. Mason; *PROFILE* No. 98
Hawker Audax & Hardy, F. Mason; *PROFILE* No. 140
De Havilland 10, J.M. Bruce; *PROFILE* No. 145
Bristol F2B Fighters, 1918-32, C. Bowyer; *PROFILE* No. 237
De Havilland 9A, 1918-30; C. Bowyer; *PROFILE* No. 248
Vickers Wellesley, N. Barfield; *PROFILE* No. 256

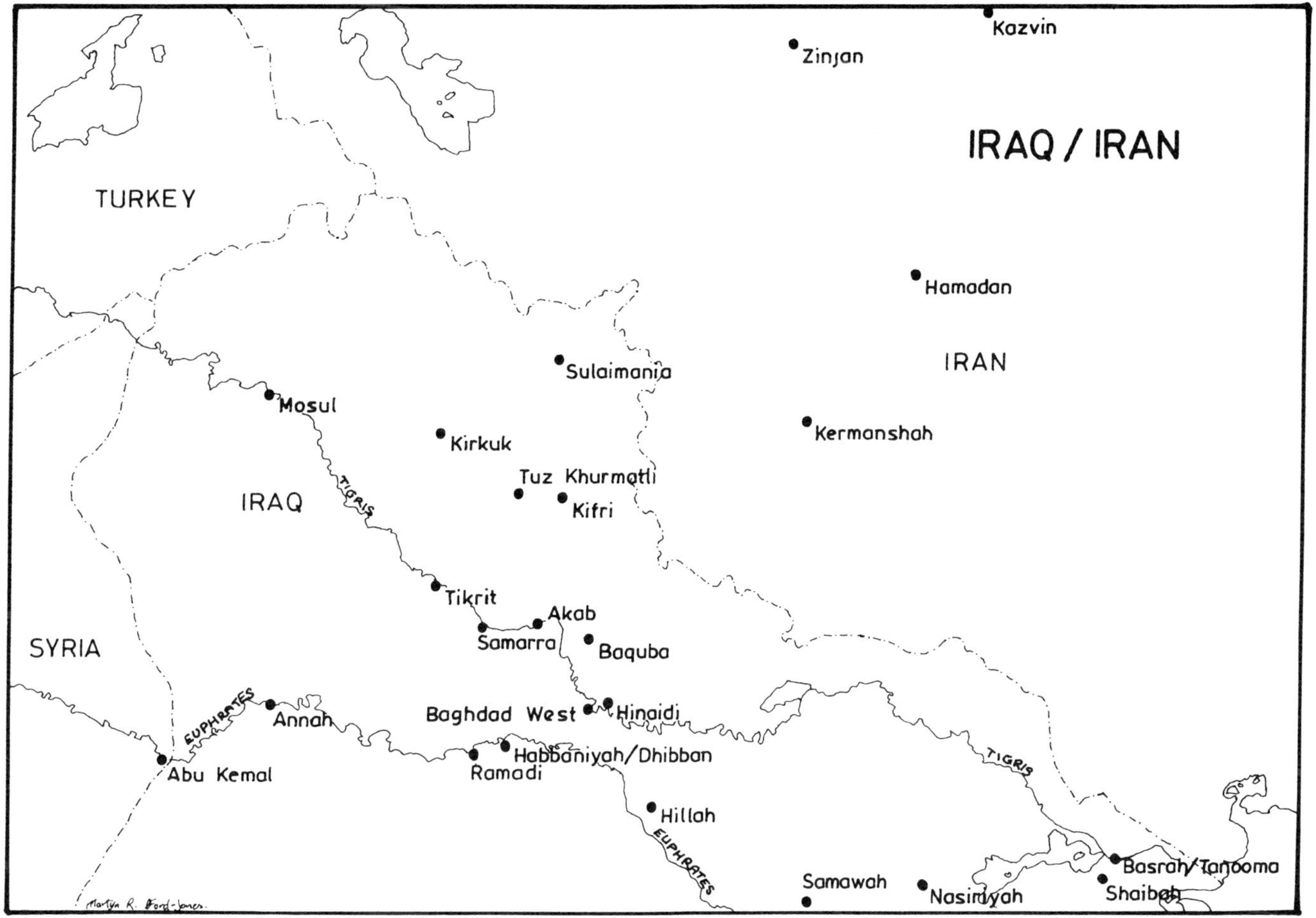
IRAQ / IRAN
Kazvin
Zinjan
TURKEY
Hamadan
IRAN
Sulaimania
Mosul
Kirkuk
Kermanshah
Tuz Khurmatli
Kifri
IRAQ
TIGRIS
Tikrit
Akab
Samarra
Baquba
SYRIA
EUPHRATES
Annah
Baghdad West
Hinaidi
Habbaniyah/Dhibban
Ramadi
Abu Kemal
Hillah
EUPHRATES
TIGRIS
Samawah
Nasiriyah
Basrah/Tanooma
Shaibah

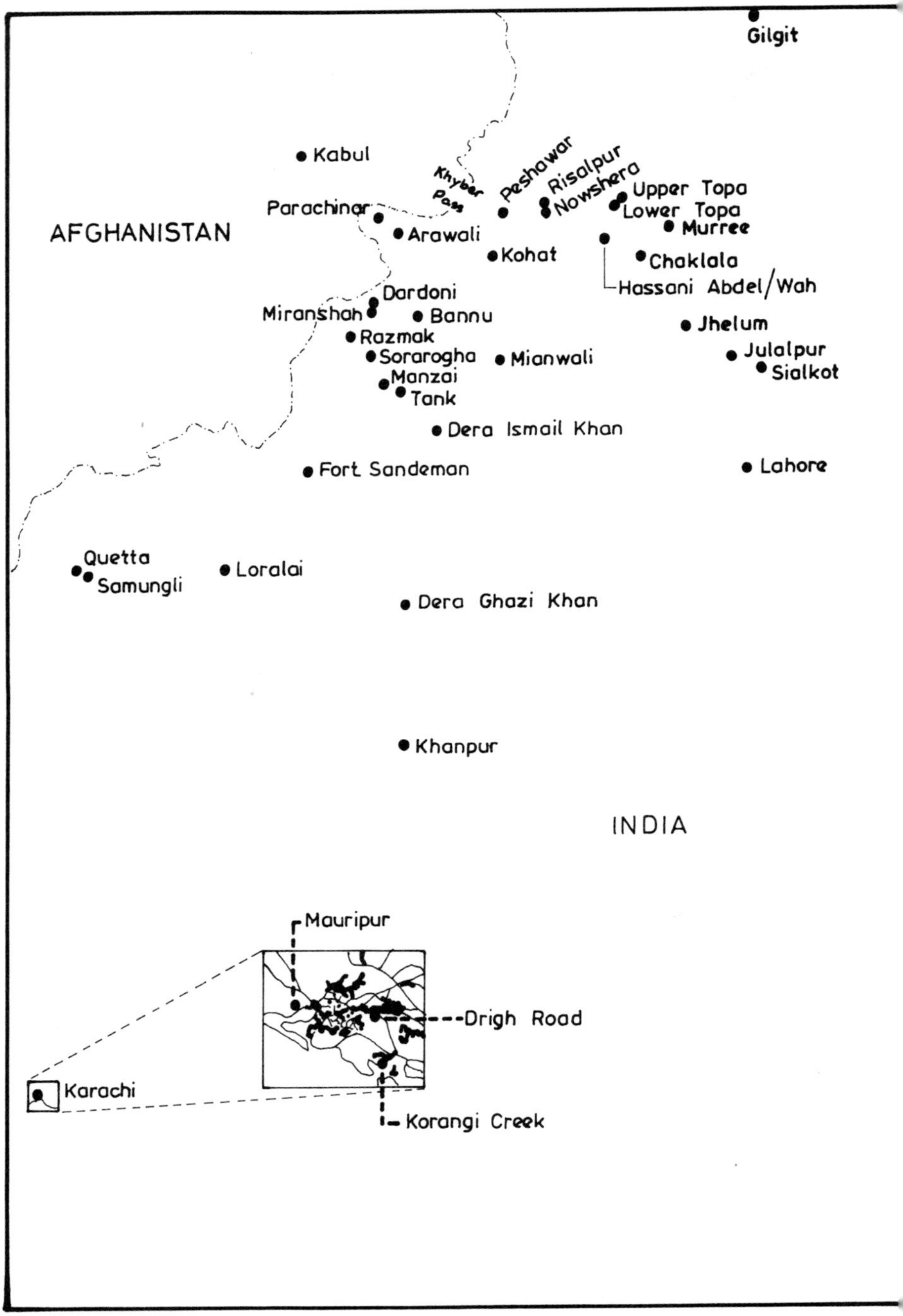
Gilgit
Kabul
Khyber Pass
Peshawar
Risalpur
Nowshera
Upper Topa
Lower Topa
Murree
Parachinar
AFGHANISTAN
Arawali
Kohat
Chaklala
Hassani Abdel/Wah
Dardoni
Miranshah
Bannu
Jhelum
Razmak
Sorarogha
Mianwali
Julalpur
Sialkot
Manzai
Tank
Dera Ismail Khan
Fort Sandeman
Lahore
Quetta
Samungli
Loralai
Dera Ghazi Khan
Khanpur
INDIA
Mauripur
Drigh Road
Karachi
Korangi Creek

NORTH WEST
INDIA
0 10 30 60 120
naut miles
CHINA
Ambala
Delhi
Palam
NEPAL
Cawnpore/Kanpur
Jubbulpore
Martyn R. Ford-Jones

Index

Names

Places

Squadrons